3rd International Conference on Geological, Geographical, Aerospace and Earth Sciences (AeroEarth 2015)

IOP Conference Series: Earth and Environmental Science
Volume 30

Jakarta, Indonesia
26 – 27 September 2015

Editor:

Ford Lumban Gaol

ISBN: 978-1-5108-2083-8
ISSN: 1755-1307

Printed from e-media with permission by:

Curran Associates, Inc.
57 Morehouse Lane
Red Hook, NY 12571

Some format issues inherent in the e-media version may also appear in this print version.

Printed by Curran Associates, Inc. (2016)

For permission requests, please contact the Institute of Physics
at the address below.

Institute of Physics
Dirac House, Temple Back
Bristol BS1 6BE UK

Phone: 44 1 17 929 7481
Fax: 44 1 17 920 0979

techtracking@iop.org

Additional copies of this publication are available from:

Curran Associates, Inc.
57 Morehouse Lane
Red Hook, NY 12571 USA
Phone: 845-758-0400
Fax: 845-758-2634
Email: curran@proceedings.com
Web: www.proceedings.com

Table of contents

Volume 30

3rd International Conference on Geological, Geographical, Aerospace and Earth Science 2015 (AeroEarth 2015)
26–27 September 2015, Jakarta, Indonesia

Accepted papers received: 4 December 2015
Published online: 5 February 2016

Papers

Preface

The 3rd International Conferences on Geological, Geographical, Aerospaces and Earth Sciences 2015 (AeroEarth 2015), was held at The DoubleTree Hilton, Jakarta, Indonesia during 26 – 27 September 2015.

The 1st AeoroEarth was held succefully in Jakarta in 2013. The success continued to The 2nd AeroEarth 2014 that was held in Kuta Bali, Indonesia. The publications were published by EES IOP in http://iopscience.iop.org/1755-1315/19/1 and http://iopscience.iop.org/1755-1315/23/1 respectively.

The AeroEarth 2015 conference aims to bring together researchers, engineers and scientists from around the world.
Through research and development, Earth’s scientists have the power to preserve the planet's different resource domains by providing expert opinion and information about the forces which make life possible on Earth.

The theme of AeroEarth 2015 is "Earth and Aerospace Sciences : Challenges and Opportunities"

Earth provides resources and the exact conditions to make life possible. However, with the advent of technology and industrialization, the Earth's resources are being pushed to the brink of depletion. Non-sustainable industrial practices are not only endangering the supply of the Earth's natural resources, but are also putting burden on life itself by bringing about pollution and climate change.

A major role of earth science scholars is to examine the delicate balance between the Earth's resources and the growing demands of industrialization. Through research and development, earth scientists have the power to preserve the planet's different resource domains by providing expert opinion and information about the forces which make life possible on Earth.

We would like to express our sincere gratitude to all in the Technical Program Committee who have reviewed the papers and developed a very interesting Conference Program as well as the invited and plenary speakers.

This year, we received 78 papers and after rigorous review, 18 papers were accepted.

IOP Conf. Series: Earth and Environmental Science **30** (2016) 011001 doi:10.1088/1755-1315/30/1/011001

The participants come from 8 countries. There are 4 (four) Parallel Sessions and 3 (Three) Keynote Speakers.

It is an honour to present this volume of *IOP Conference Series: Earth and Environmental Science (EES)* and we deeply thank the authors for their enthusiastic and high-grade contribution.
Finally, we would like to thank the conference chairmen, the members of the steering committee, the organizing committee, the organizing secretariat and the financial support from the conference sponsors that allowed the success of AeroEarth 2015.

The Editors of the AeroEarth 2015 Proceedings Editors

Dr. Ford Lumban Gaol
Bina Nusantar University

IOP Conf. Series: Earth and Environmental Science **30** (2016) 011001 doi:10.1088/1755-1315/30/1/011001

IOP Conf. Series: Earth and Environmental Science **30** (2016) 011001 doi:10.1088/1755-1315/30/1/011001

(TelSaTech 2015)
www.telsatech.org
(AeroEarth 2015)
www.aeroearth.org

IOP Conf. Series: Earth and Environmental Science **30** (2016) 011001 doi:10.1088/1755-1315/30/1/011001

IOP Conf. Series: Earth and Environmental Science **30** (2016) 011001 doi:10.1088/1755-1315/30/1/011001

IOP Conf. Series: Earth and Environmental Science **30** (2016) 011002 doi:10.1088/1755-1315/30/1/011002

Keynote Speakers

1. DR CAO, KAI
National University of Singapore

Title: Spatial Optimization for Sustainable Land Use Planning

Sustainability often represents a primary goal for land use planning. Comprehensive sustainability in land use planning can be defined as a long-term balance between economic development, environmental protection, efficient resource use, and social equity. To pursue this balance spatially and intelligently, a series of loose coupling models and technologies from the perspectives of spatial optimization and GIS have been developed and employed to create the optimal/near-optimal land use planning scenarios to support the sustainable land use planning. The proposed framework could also be flexibly adapted to other spatial optimization problems.

Speaker Bio

Dr. Kai Cao holds a Ph.D. degree in Geography from the Chinese University of Hong Kong in Hong Kong and obtained his B.S. and M.Phil. degrees in Geography from Nanjing University in China. Prior to joining Department of Geography at NUS, he had also worked in Center for Geographic Analysis at Harvard University, Department of Geography at University of Illinois at Urbana–Champaign, and World History Center at University of Pittsburgh respectively. His current research interests center on GIScience and its applications, especially on the topics of spatial simulation and optimization, cyberinfrastructure and geocomputation, land use planning support and urban mobility.

IOP Conf. Series: Earth and Environmental Science **30** (2016) 011002 doi:10.1088/1755-1315/30/1/011002

2. Dr Boni Pudjianto

Deputy Director for ICT Empowerment in Urban Area

Ministry of Communication and Information Technology, Republic of Indonesia

Title : The Geo-Spatial Data Role in Indonesia Development

Speaker Bio

Bonifasius Wahyu Pudjianto is currently serving as the Deputy Director for ICT Empowerment in Urban Area at Directorate General ICT Application, Ministry of ICT, The Republic of Indonesia. His current duty mostly deals with empowering community and people to utilize ICT for enhancing their value as well as to reduce digital divide.

Previously, he was also responsible for e-Government implementation, particularly on E-Government Application. He has received recognition on government projects related to design and implementation of safe and secure internet programs, establishing the Presidential Instruction on e-Government, developing several e-Government standards, procedure and guidelines, initiating and developing a national e-procurement system, and serving as overseer on an ICT Learning Project for Yogyakarta Province. He is also active in several international co-operations and a member of International working group such as ApecTel WG on E-Government, WSIS (Indonesia committee member), ITU, Open e-Governance Forum Korea Indonesia Cooperation projects, and many others.

He obtained his Doctoral Degree in IT Technology Program at KAIST University, South Korea and finished his study as Master of Production Process Engineering at Bandung Institute of Technology (ITB) as Cum Laude Student for his Master's degree. His bachelor's degree was obtained from Bandung Institute of Technology (ITB) in Mechanical Engineer in 1992.

IOP Conf. Series: Earth and Environmental Science **30** (2016) 011002 doi:10.1088/1755-1315/30/1/011002

.

3. Dr Nawaz Muhammad
National University of Singapore

Title : **Earth Science Teaching, Learning and Research: Challenges and Opportunities in Low Resource Settings**

Earth science teaching, learning and research has gained significant attention in recent years. Earth science today is a space-age science making extensive use of remote sensing data, high performance computing, spatial data base management, analysis, visualization, dissemination and sharing of spatial data. Geographic Information System (GIS) and Remote Sensing (RS) software are an integral part of earth science teaching, learning and research. Despite the demonstrated utility of commercial/proprietary tools such as ArcGIS and ERDAS IMAGINE for teaching and learning in earth science there are problems in accessing this software in low resource settings. Proprietary GIS and RS software are expensive and often more complex than required. A new generation of free and open source software are now available which have potential to positively impact geoscience teaching and learning.
The goal of this keynote presentation is to assist teachers and researchers at institutes with limited budgets to understand and appreciate the potential of RS and GIS free/open source software in earth science education and research. This research explores the possibilities using free/open source software, provides a list of software and elaborates their uses in earth science teaching, learning and research. General analysis reveals that for all the categories of commercial software used in geoscience teaching, learning and research, a suitable free/open source alternative is available. These free tools are particularly useful for handling low quality spatial data and are therefore suitable for use in institutes with limited budget.

IOP Conf. Series: Earth and Environmental Science **30** (2016) 011002 doi:10.1088/1755-1315/30/1/011002

Speaker Bio

Dr Nawaz Muhammad is geomorphologist with expertise in Geographic Information System (GIS) and Decision Support Systems (DSS).
Since 1993, Dr Nawaz has served in several different universities including Punjab University, University of Twente in Netherlands and Charles Darwin University in Australia. He has vast experience and knowledge in design, development, delivery, coordination and teaching of courses related to Physical Geography, GIS and Environmental Science to undergraduate, graduate and postgraduate level students.
Dr Nawaz has over 20 years of research over a broad range of issues relating to catchment management and sustainable development. Before joining NUS, Dr. Nawaz worked as Lecturer in Environment Science eLearning at Charles Darwin University, Australia and before this he has worked as Associate Head/Assistant Professor at GIS Centre, University of the Punjab. He has conducted multi-disciplinary research in several catchments including Darwin Harbour and Daly River catchment, Australia, Mangla Reservoir catchment, Pakistan and Saddang Watershed, TanaToraja, South Sulawesi, Indonesia.
Dr Nawaz's expertise includes sediment source identification, multicriteria evaluation, and GIS applications in decision making.
Dr Nawaz has been awarded a doctorate degree in Earth and Environmental Science from the Charles Darwin University, Australia, Master of Science in Geoinformatics from the International Institute of Geo-Information Science and Earth Observation (ITC) – University of Twenty, the Netherlands, Master of Science in Geography from the University of the Punjab, Pakistan and Bachelors of Science in Geography, Chemistry and Zoology from the Bahauddin Zakariya University, Pakistan.

IOP Conf. Series: Earth and Environmental Science **30** (2016) 011003 doi:10.1088/1755-1315/30/1/011003

Peer review statement

All papers published in this volume of *IOP Conference Series: Earth and Environmental Science* have been peer reviewed through processes administered by the proceedings Editors. Reviews were conducted by expert referees to the professional and scientific standards expected of a proceedings journal published by IOP Publishing.

IOP Conf. Series: Earth and Environmental Science **30** (2016) 012001 doi:10.1088/1755-1315/30/1/012001

Effect of Frequency and Migration Aperture on Seismic Diffraction Imaging

Y. Bashir[1], D. P. Ghosh[1], S.Y. Moussavi Alashloo[1] and C. W. Sum[1]
[1] Centre for Seismic Imaging (CSI), Department of Petroleum Geoscience, Universiti Teknologi PETRONAS, Tronoh, Perak, Malaysia

E-mail: Yasir.bashir@live.com

Abstract. Conventional processing and migration frequently give successful results in using specular reflections to estimate the subsurface geometry and strength of continuous reflector geology. However, the correct interpretation of the true geological gaps, such as fault, fracture, karsts and pinch-outs, is one of the main objectives in seismic data processing and interpretation. In regular processing/migration sequence the diffraction response is suppressed because of the lack of choosing the right migration aperture. Kirchhoff migration is a tool to represent the seismic data as a summation of diffraction hyperbolas governed by the velocities at their apex. In this paper, we have investigated two different velocity models to show the effects of different frequencies and aperture size. We used the diffraction-based and data oriented approach that is dependent on the migration aperture from a low to high aperture to properly image the section. We have done the error analysis between the un-imaged and imaged section after processing and observed that the low aperture can give the undesired result for sharp edges. For the same model, we have applied different frequencies to show the effect of frequencies on seismic Imaging and migration.

1. Introduction

Seismic data Acquisition and imaging are inverse processes in exploration geophysics. Acquisition belongs to the propagation of the wave to get the true geology but unfortunately that wave is affected by many factors in the subsurface and also depends on the property of the wave. One of the factors here is that the initial frequency of our seismic wave is affected by many factors during the wave propagation. This is because high frequency wave give us high resolution but less depth of penetration and a low frequency one have high depth of penetration but less resolution[1]. A relation between the frequency, penetration, resolution, wavelength, and diffraction hyperbola can be defined as.

High Frequency →Less depth of Penetration → High Resolution
Low Frequency →Deep depth of Penetration → Low Resolution

Frequency is inversely proportional to the wavelength.

High Frequency →Small Wavelength

[1] Yasir Bashir (Yasir.bashir@live.com)

Low Frequency → Large Wavelength

$$f \propto \frac{1}{\lambda}$$

High Frequency → Less diffraction
Low Frequency → high diffraction

Where f is the frequency and λ is the wave-length. Resolution will be $\lambda/4$, which means that if we are using a high frequency that will illuminate a small object; but, increasing the frequency will also have an effect on the depth of penetration. So, we can define the frequency according to our objective. Seismic methods use low frequencies. Thus, they have low resolution but high depth of penetration. This suits us for imaging the earth. The thinnest layers are in the order of several metres and may be a few kilometres deep. The seismic waves can propagate down to such depths and resolve these layers[1].

There are different phenomena in wave propagation, e.g., Refraction, reflection, diffraction and transmission. Diffraction hyperbolic patterns occur frequently in a seismic section, and their existence is usually taken as evidence of abrupt discontinuities in subsurface reflector geometry[2]. The curvature of the diffraction hyperbola is dependent on the different depths and velocities of the layer. As a high frequency will produce less diffraction then a low frequency and with the depth increase, the diffraction hyperbola will spread out more and the apex will be gentle[3]. Kirchhoff migration sums the amplitude along the diffraction hyperbola and brings all the energy at the apex. The aperture size is used for the summation of the hyperbola amplitude and this is very important parameter in migration.

Seismic data processing and imaging is a tool to image the wave propagation travel time and illuminate the actual geology. This paper deals with increasing the understanding of Kirchhoff migration, precisely on the different frequencies and aperture sizes on seismic imaging.

2. Methodology

2.1. Theory and Method

Diffraction theory states that a diffraction hyperbola extends to the infinite time and distance. In practice, we have to deal with truncated hyperbolic summation paths[4]. Spatially, the extent of the actual summation path span or aperture width is measured in terms of the number of traces the hyperbolic path spans[5]. In general theory, diffraction hyperbola will be increase with time and distance and in exercise we have to sum the travel time which is calculated on different position. Aperture width can be measure in term of the traces numbers per hyperbola[5]. The equation describing the hyperbola is given as

$$t^2(x) = t^2(0) + \frac{4x^2}{V^2}$$

Where x is the distance in (m) from the output trace to that of the input trace location and t (x) is the input time computed and V is the velocity in (m/sec).
Kirchhoff migration is the best method to describe the kinematics. The source and receiver compute the energy from the reflection point in time at each point diffractor, which is a point from the reflector to the receiver and is recorded in time[6]. Technically, the constant velocity Kirchhoff migration programme needs each input un-image data to visit each output data traces with in a migration aperture exactly once. This parameter allows one simulate, accurately the number of additions that will be performed in the migration algorithm[6].

IOP Conf. Series: Earth and Environmental Science **30** (2016) 012001 doi:10.1088/1755-1315/30/1/012001

2.2. Velocity Model

In this study, we have considered two geological Models called hole in reflection. Model-1 had a hole width of 100 m (Figure 1) and Model-2 had a Hole width of 300 m (Figure 2). For convenience, we chose the constant velocity and variable density model. Having a velocity of 2500 *m/sec*, density of the1[st] layer was 2.2 Kg/m^3 and 2.7 Kg/m^3, respectively, and the total depth of the model was 2000 *m*.

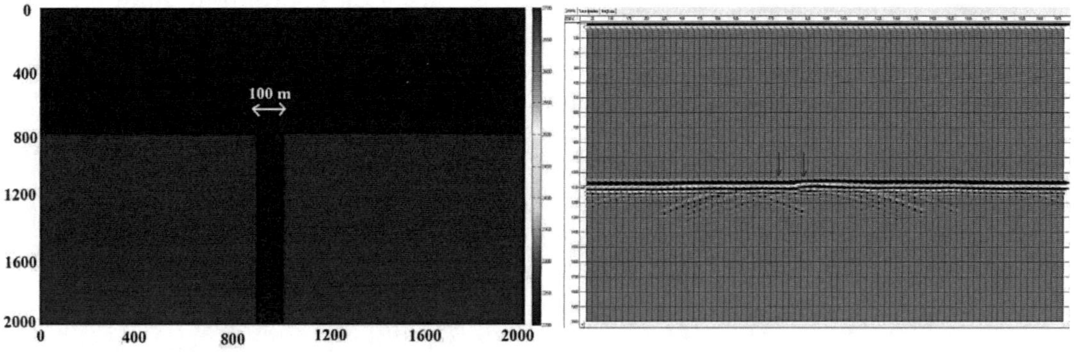

Figure 1: Velocity model-1 with a hole width of 100 m and synthetic seismic gather.

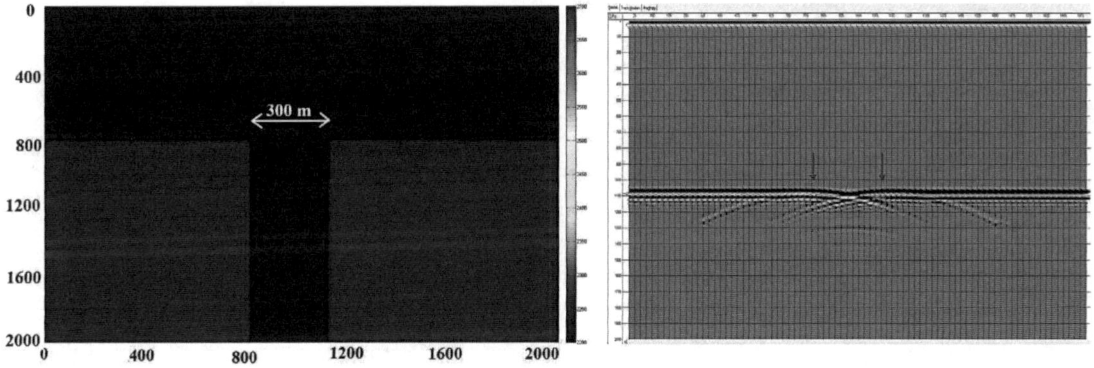

Figure 2: Velocity model-2 with a hole width of 300 m and synthetic seismic gather.

2.3. Seismic Data

Over the velocity model, 2D synthetic data were acquired using the finite-difference wave equation technique. The common shot section was acquired along a 2000 meter seismic line. In this model, we used the zero-offset survey design, and the distance between each shot was 10 meters and the sampling interval was 2 Milliseconds.

Seismic data was acquired in two different frequencies, 30 Hz and 80 Hz. As stated in the theory, the diffraction hyperbolic effect was more in the low frequency than the high frequency as shown in Figures 3-5.

2.4. Seismic Migration

Implementation of the Kirchhoff migration was accomplished with the use of asymptotic calculation that is effective for large values of frequency and time. In this paper, we have applied the Kirchhoff migration on low frequency and high frequency.

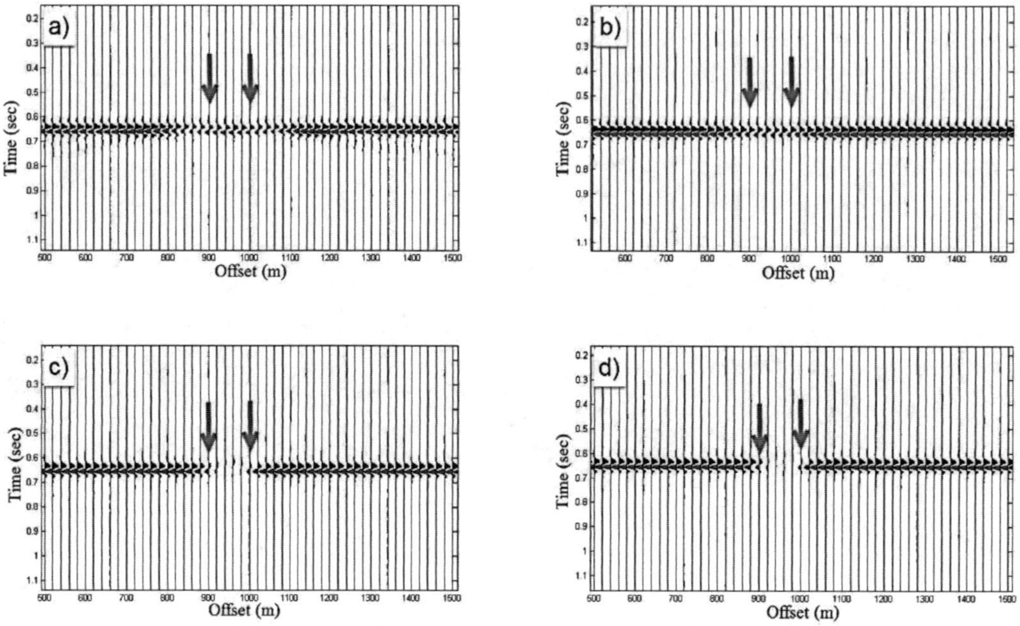

Figure 3: Hole in reflection model-1 with a dominant frequency 30 Hz and a Hole width of 100m. (a) Zero-offset Synthetic Seismic gathers on Model-1, (b) (c) and (d) are traces after the application of the Kirchhoff Migration on different aperture sizes of 100, 300 and 500m, respectively.

3. Results and Discussion

The example shown in the synthetic seismic gather illustrates a pitfall to avoid with the use of migration aperture. Figure 3 illustrates the different aperture sizes affecting the diffraction hyperbola to be imaged using the Kirchhoff summation algorithm. It can be observed by visual interpretation during processing how much aperture size was necessary to image the proper geological feature. For some of the case we did not need a large migration aperture because if the aperture size was large then the processing time would be high and computational power would also increase to bring all the energy from the flanks to the apex. Generally, if the depth increases then the velocity will also increase; so, according to the theory, if the velocity is high, then the curvature of the hyperbola will be less. During processing, Kirchhoff actually sums the energy from the flanks and combines it on the apex; but, in the case of high depth, we could not sum it correctly because of not knowing which energy belonged to the hyperbola and which reflection amplitude it was. So, for a high velocity area, we still had difficulties to image the diffraction hyperbola. In this situation, we need to refer to the wave-equation migration that is in two domains, One-way and two-way.

For Model-1 and 2, we generated a synthetic gather in two different frequencies to show the effect on the seismic imaging. Low frequency data was imaged on a larger aperture size than the high frequency data as shown in Figures 3-6.

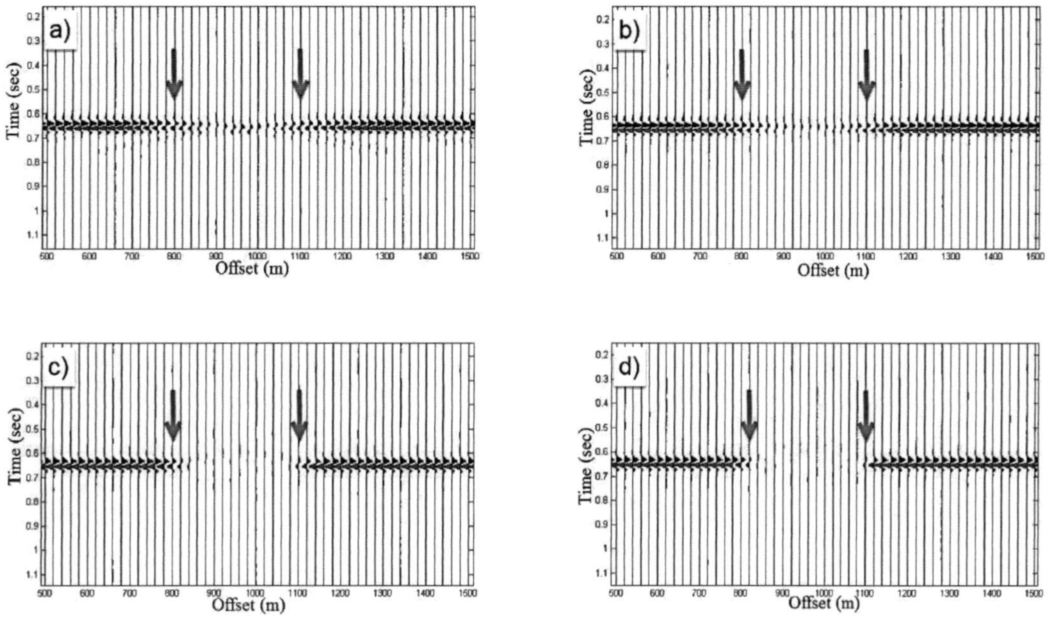

Figure 4: Hole in reflection model with a dominant frequency of 30 Hz and Hole width of 300m. (a) Un-Image section, (b), (c) and (e) are migrated sections on different aperture sizes, 100, 300 and 500 m, respectively.

Figures 3 and 4 illustrate the low frequency data on Model-1 and Model-2, respectively, which had 100 m and 300 m hole widths. 3a and 4a show the un-Image zero-offset seismic gather, 3b and 4b are after the application of the Kirchhoff migration using the aperture size of 100 m; but, we can see an error because of choosing the small aperture. For sections 3b and 4b, the sharp edges are not illuminated until we increased the migration aperture until 400 m as shown in Figures 3d and 4d.

Here, if we imagine the real earth model, generally increasing the depth, the frequencies will also effected by earth materials wave and the high frequencies will be absorbed. So, in deep sections we have only low frequencies; these low frequencies will give us high diffraction responses as stated above. In actuality, we have to use a larger aperture for a deep diffraction response summation. For our model, we did not have that depth, so we used a small aperture that was enough to image the diffractions.

Figures 5 and 6 illustrate the high frequency synthetic seismic gather on Model-1 and Model-2, the hole width was 100 m and 300 m, respectively. 5a and 6a are zero-offset un-Image seismic gathers in which we cannot interpret the fault, which is described by the blue colour arrow. 5b and 5c are after application of the Kirchhoff migration using the aperture size of 100 m. 5c and 6c are also after the migration using aperture length of 300 m. We have observed that in a high frequency, the migration aperture of 300 m is enough to image the section rather than the low frequency as shown in Figures 3 and 4. High frequencies are good for the resolution purpose but for the deep events, we have to use a low frequency that can be imaged by increasing the migration aperture size.

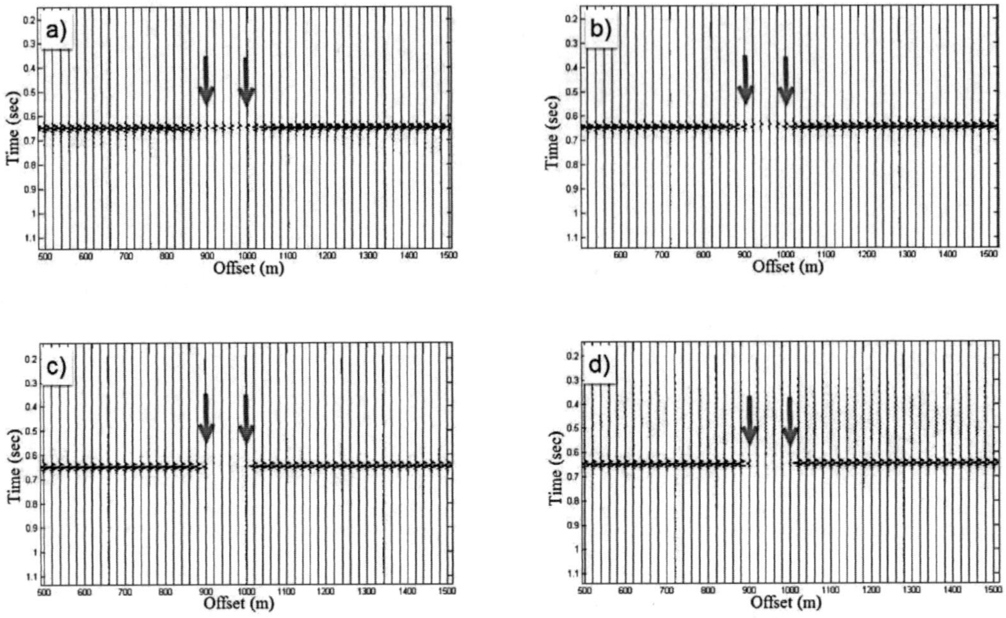

Figure 5: Shows traces with a dominant frequency of 80 Hz and a Hole width of 100m. (a) Zero-offset Synthetic Seismic gathers on Model-1, (b), (c) and (d) are traces after application of the Kirchhoff Migration on different aperture sizes.

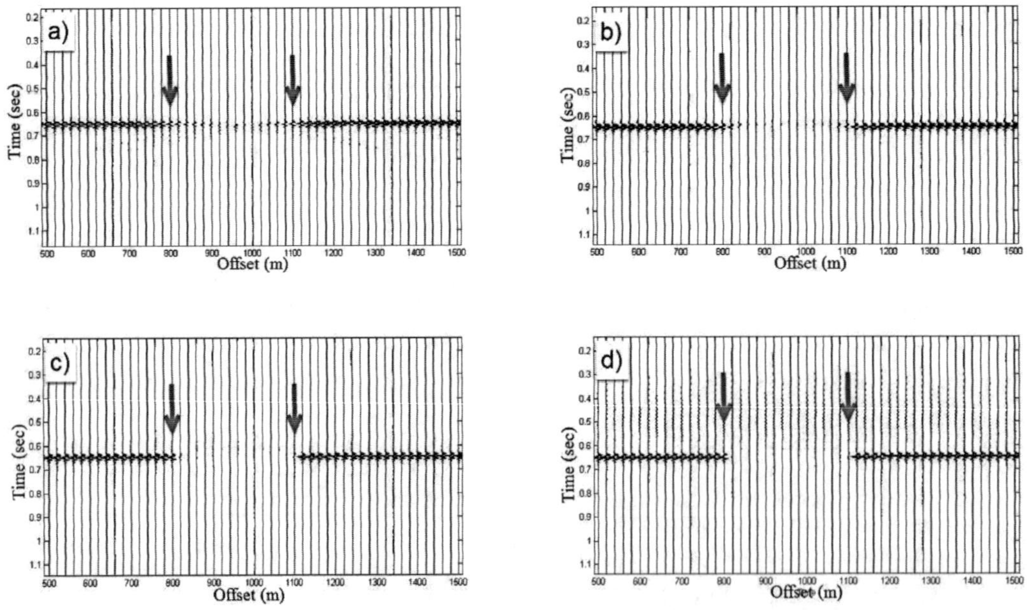

Figure 6: Hole in reflection synthetic gather with a dominant frequency 80 Hz and a Hole width of 300m. (a) Un-image section, (b) Migrated section with aperture of 100 m, (c) Aperture of 300 m and (d) Aperture of 500 m.

Conclusion

We have demonstrated that the traditional processing/migration regard the diffractive wave field as noise and ignore the structural events which are conveyed by the diffraction. During processing (NMO/DMO, multiple attenuation), events that are not originating from a flat and robust reflector are usually filtered out. These results make images only the highlight reflectors and obscure non-reflecting structures such as faults, edges, pinch-outs and small scattering objects. The main challenges of the diffraction analysis are that the diffractions are usually tangent to the reflection and phase change is 180 degrees with the exponential energy decay. Seismic imaging using the Kirchhoff migration algorithm has been applied to sum the diffraction response from the edges of the reflector and using the different aperture sizes to image the diffractions. We have Optimised that data which had high frequencies that needed the small aperture length, and for low frequencies a small migration aperture.
Results include the error analysis because of wrong selection of migration aperture for different frequencies in seismic migration. A good quality migrated image requires a migration aperture with an optimal size. Disproportionately small apertures cause destruction of the steeply dipping events and edges of the reflector. More importantly, large apertures will degrade the migration quality in a poor signal-to-noise ratio and increase the cost and computational power without quality improvement.

Acknowledgments

We are thankful to Universiti Teknologi PETRONAS (UTP) and the Centre for Seismic Imaging (CSI) to provide the facility for this research. Y. Bashir is thankful to PETRONAS for supporting him financially. In this work, we have used MatLab and other tools like CREWES.

Reference

1. Khan, K.A., *SAARC Workshop on Geophysical Techniques for Exploration of Natural Resources.* 2010(2010): p. 106.
2. Hilterman, F., *Amplitudes of seismic waves; a quick look.* Geophysics, 1975. **40**(5): p. 745-762.
3. Y. Bashir, D.G., C.W.Sum, *Detection of Fault and Fracture using Seismic Diffraction and Behavior of Diffraction Hyperbola with Velocity and Time.* EAGE 2015: p. 6.
4. Rastogi, R., S. Yerneni, and S. Phadke. *Aperture width selection criterion in Kirchhoff migration.* 2000. Citeseer.
5. Yilmaz, O., *Seismic Data Processing* Geophysic, 1987. **2**(Society of Exploration Geophysicists).
6. Gray, S.H., et al., *Seismic migration problems and solutions.* Geophysic, 00/2001. **66**(5): p. 1622.

Interpretation of Gas Seepage on Seismic Data: Example from Malaysian offshore

S.N. Fathiyah Jamaludin[1], Abdul Halim Abdul Latiff [2], Askury A.Kadir[3]

[1,2,3,] Geosciences Department,
Universiti Teknologi PETRONAS,
Bandar Seri Iskandar, Ipoh,
32610 Malaysia

E-mail: fathiyah.jamaludin@petronas.com.my[1]

Abstract: Gas seepage is one of the geological hazard in the subsurface exploration. Migration of gas usually originated from the deeper reservoir due to poor sealing properties of the cap-rock as well as existence of faults or fractures within the reservoir. In order for us to get familiar with the existence of gas seepage on seismic data, we need to know their characteristics and how do they look like. This paper present several features of gas seepages as viewed on seismic data and how to straight away recognize them at the first time viewing the data. Detailed observation of the seismic data affected by gas leakage shows that the top of the gas leakage always exhibit bright amplitude with notable wipe-out zone within the gas chamber. The seismic reflections within the gas seepage also gives chaotic reflection and give difficulties in interpreting the structure within the gas seepage. Use of seismic attributes in characterizing the gas seepage is also presented in this paper as an aid to improve the interpretation in the area/zone affected by the seepage.

Keywords: Gas seepage, seismic interpretation, seismic attributes, seismic anomalies, wipe-out zone

1. Introduction

Hydrocarbon leakage particularly gas seepage from the deeper reservoir is a common occurrence in most of the hydrocarbon producing basins all over the world such as in the North Sea Basins [1], Timor Sea Basin [2], Malay Basin [3] and Australian Basin [4]. The existence of hydrocarbon leakage in the rock formations had become an indication that there is a larger hydrocarbon reservoir sitting quietly underneath all the formations, waiting to be explored. However, before we can reach the bigger reservoir underneath this seepage, we have to be extra careful in managing the field with the hydrocarbon seepage. This is because, the seepage is also considered as a geological hazard especially while drilling the wells and placement of the rig, Apart from that, this hydrocarbon seepage is also an obstacle for the geophysicists to correctly process and interpret the seismic data due to the acoustic blanking appearance created by the seepage. Hydrocarbon seepage may be studied for several different reasons. In this paper, the authors present the characteristics and representation of hydrocarbon leakage, with emphasis on the gas seepage as observed on the seismic data. The example of seismic data used in this paper is from a producing hydrocarbon field from clastic reservoir in the Malay Basin, offshore Malaysia.

[1] Corresponding author.

2. Representation of Gas Seepages on Seismic Data

Gas seepage on seismic data is categorized as a seismic anomaly, which means it brings interruption to the data as shown in Figure 1. According to [5], poor illumination of seismic data with gas seepages is a result of scattering, attenuation and decrease in velocity of compressional waves (P-waves) passing through the gas-effected zone. Løseth, et al. [6] had introduced a simple three steps workflow to interpret any hydrocarbon leakage on the seismic data, which are observed, described and finally mapped the leakages. Observation and description of the gas seepage on the seismic data required the interpreters to know the characteristics of these seepages. The gas anomalies on seismic data are variable in sizes and shapes but usually show the same seismic characteristics such as chaotic reflection, poor amplitude within the gas seepages and low velocity zone.

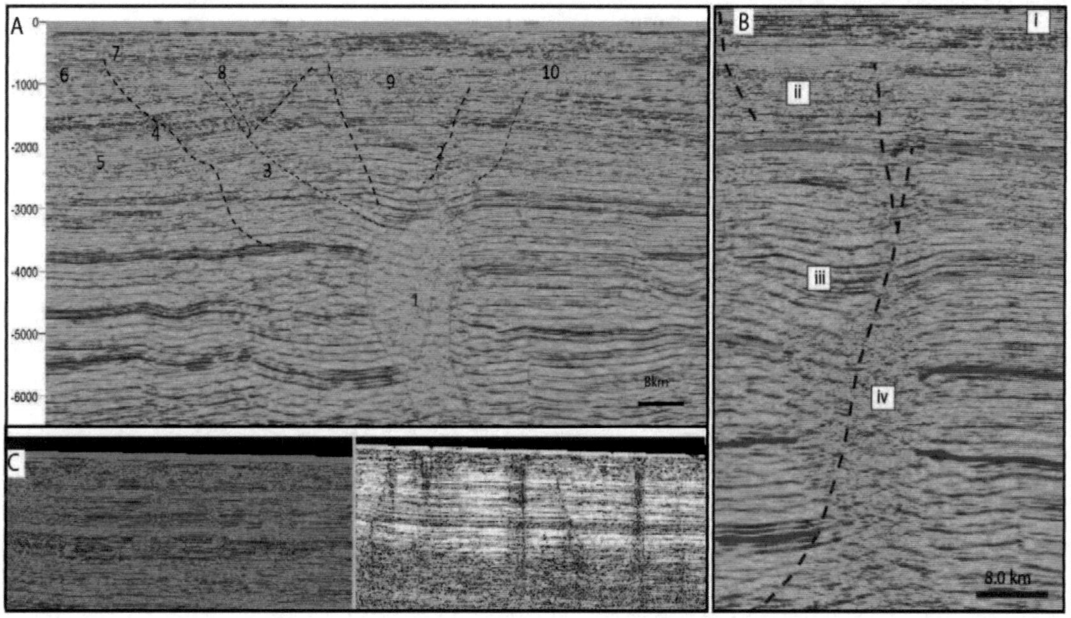

Figure 1: A) Labelling of Gas seepages (1-10) observed on the seismic data used in this study with Seepage-1 being the largest seepage and blanking the area of about 10-15km wide and 8.0km deep. A major fault is interpreted to be responsible for the migration of the gas from deeper reservoir based on the notable displacements of the horizons. Seepage 2-10 are randomly distribute all over the seismic section particularly on the shallow part of the data with some interactions with faults. B) Zoomed Section from Figure 1A (rectangle) summarized the characteristics of gas seepage on seismic data. i: High Amplitude (Direct Hydrocarbon Indicator (DHI)) at top of the seepage; ii: Chaotic, blurry with sand papery look that dimmed the reflection; iii: Time-sag (push down effect); iv: Acoustic Blanking (vertical wipe-out zone). C) Shallow section of the data, viewed on original amplitude and variance attributes. Variance attribute can clearly locate the location of the faults and gas seepages associated with it.

In this work, we had interpreted ten individual gas seepages (Figure 1) that are variable in shapes and sizes but show typical seismic features such as *wipe-out zone (acoustic blanking)* for larger gas seepage, *vertical dimmed zone, high amplitude (bright spot/flat spot) at the top* of the seepages, *low continuity reflector* within the effected zone, *chaotic reflection pattern with sand-paper look* and *local depression or time-sag (push down effect).* The gas seepage can be divided into two groups, either they are migrated seepage through structural deformation in the rock formation or localized gas seepage. The summary of our interpretation is presented in Table 1.

Journal of Physics: Conference Series **660** (2015) 012002 doi:10.1088/1742-6596/660/1/012002

Gas that seeps out from the reservoir can create physical deformation in the sedimentary layering. This usually takes place in soft-sediments and mostly associated with formation of fractures. The physical deformation can be characterized as either *pockmarks or craters* (Figure 2). Pockmarks are depression observed on the shallow sea-bed with few metres to tens of metres deep, within soft, fine-grained sea-bed sediments, produced by the outbreak of fluid or gas into the water column [7]. Craters, on the other hand are larger circular sea-bed depressions that may take up to few hundreds of metres in radius and several metres deep, and usually occurs in relation to the formation of faults and fractures**.** Pockmarks and craters are two common examples of gas seepage representation on the seismic data used in this research as presented on Figure 2.

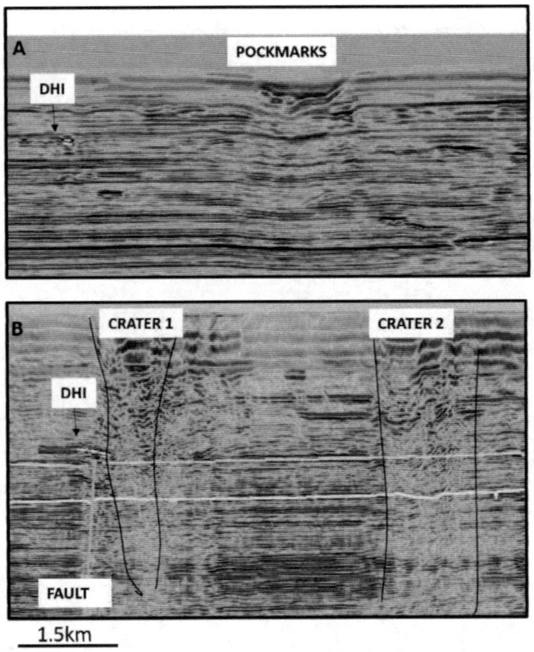

Figure 2: A) Example of sea-bed depression or pockmark. This depression is about 1.25km wide and 17m deep with circular, bowl shape structure on cross-section. B) Example of craters associated with fault at shallow section of the data. Crater 1: 1.484km deep, 700m wide. Crater 2: 1.5km deep, 1.2km wide. Depth is measured from the interpreted surface that originated the gas to the top (high amplitude) of the gas seepage.

Another characteristic of a gas seepage zone is the amplitude anomalies. First amplitude anomaly is the bright amplitude as *Direct Hydrocarbon Indicator (DHI)* in the cap rock of the gas seepage zone (Figure 1 and 2). The bright amplitude usually captures the attention of the interpreters if it is properly sealed. However, in the case of hydrocarbon seepage zone, this type of high amplitude anomaly is located above pressured hydrocarbon (in our case-gas) accumulations, without any structural deformation within the sealing rock. Bright Spot or DHI associated with gas seepage might be due to the continuous gas saturations below a porosity and/or permeability boundary in the rock formations as explained by Løseth, et al. [6]. Second amplitude anomaly for gas seepage zone is *acoustic blanking* in sediments affected by the gas effected area. Acoustic blanking can be said as an area where the seismic data is loss (wipe-out) due to very poor reflection and scattering of the seismic wave.

A large gas cloud originated from the deeper part of the reservoir can be clearly observed and interpreted in this study. This gas seepage is hereby labelled as Seepage-1 (Figure 1 and Table 1). It gives an acoustic blanking and had camouflaged all the seismic reflection within the gas seepage, thus resulting in difficulties in interpreting the continuity of the stratigraphic layering in this section, as well as projecting is there any structural deformation within the affected zone. Many researchers [8, 9] had proven that, to improve the interpretation in a gas seepages area, application of seismic attributes such as coherency and relative acoustic impedance are extremely useful. In this study, we had compared the original amplitude data with Variance-Coherence, Chaos and Structural Dip Attributes to delineate more information about the seepages. Each attribute's algorithm is computed differently to serve for different interpretation and imaging targets.

3. Seismic Attributes on Gas Seepage

Due to poor amplitude within the gas seepage zone, we had first viewed our seismic data using Variance-Coherence attribute that works based on the continuity of the seismic wavelet. A highly coherence reflector reflects the wavelet continuity, thus low coherency zone on the data indicate poor continuity in the wavelet. Variance-Coherence attribute had clearly imaged the location of gas craters on our seismic data as seen on Figure 1C and Figure 3. Utilization of this attribute had make it easier for us to see the origin of the gas from the deeper part. Based on Figure 3, Variance and Chaos attributes had allowed us to see the continuity of the sediment layering in the gas seepage zone.

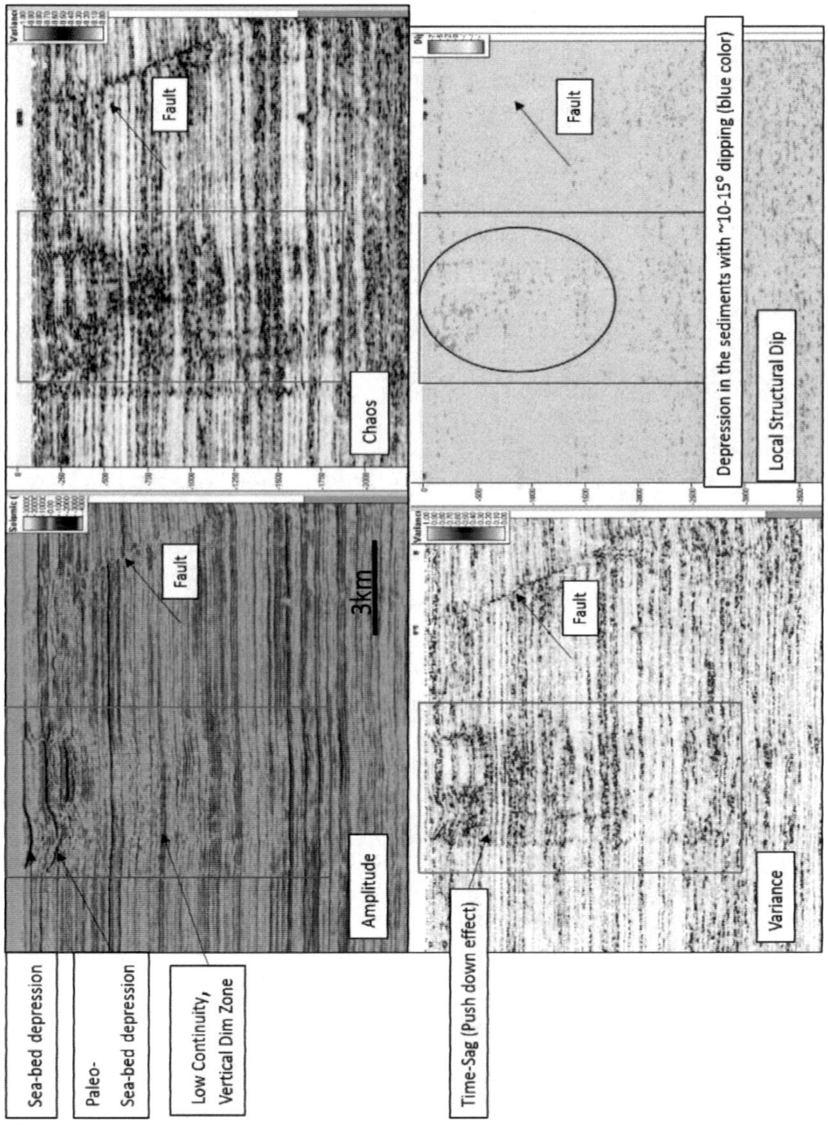

Figure 3: Gas seepage (in rectangle) on seismic data represented on different seismic attributes. The sea-bed depression can be clearly seen on the original amplitude. Chaos and Variance attributes shows disturbance and time-sag within the gas seepage with emphasis on the nearby faults. Local structural dip attribute had assist on evaluating the depression effect (10-15^{0} dip) cause by the gas seepage on the soft sediments layers.

Another effective attribute used for analysis in this study is Local Structural Dip attribute. Local Structural Dip is a powerful in capturing properties of the seismic data on the basis for compensating the value for the dipping events. Since, most of our interpreted gas seepages are related to faulting and fractures, Local Structural Dip had greatly assist us to estimate the degree of permanent depression within the sediments that is effected by the seepage which is more or less 10-15° dipping from the original horizontal bedding.

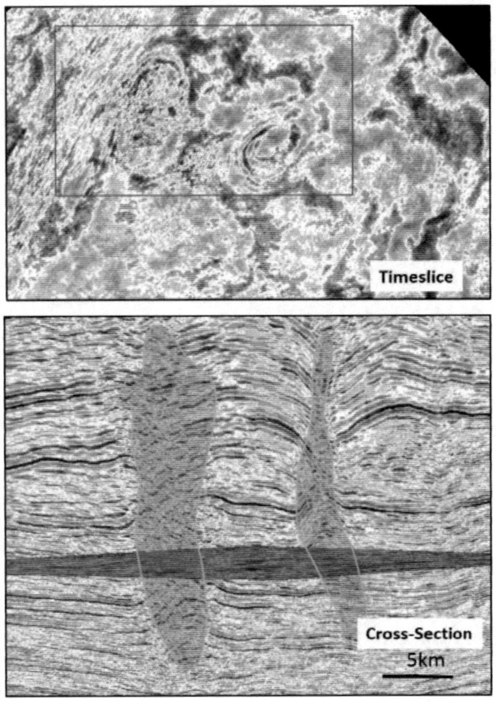

Figure 3: Seismic expression of vertical-wipe out zone caused by gas-pipe as seen on the cross section in the deeper part of the seismic. These gas-pipes look like circular depression on the time slice, which had gave a permanent disturbance in the rock formations.

In the deeper section of the seismic data, gas seepage are more prominent and easily noticeable without application of any seismic attribute, due to its larger size and loss of the amplitude. As shown in Figure 4, the gas seepages in the deeper section of our data are represented by large vertical pipes that wipe out most of the reflections behind them. This is an example of seepages form due to migrated gas from the reservoir sitting underneath. Migration of the gas from the reservoir is common due to tectonic activity in the area that created a path for the hydrocarbon to escape (usually via faults and fractures), together with poor sealing properties in the cap-rock and surrounding rock formations. If we view these vertical gas pipes on the time-slice, it will look as a circular depression which may indicate a prolong impact of gas migrating from the deeper source and voyaging within the sediments.

4. Conclusion and Future Works

Based on this study, detection of gas seepage on seismic data can be characterized based on several features. The observations include amplitude anomalies, which are poor amplitude continuity that leads to vertical wipe out zone/dimmed zone and acoustic blanking in the area covered by the gas and high amplitude anomaly at the top of the leakage zone (known as Direct Hydrocarbon Indicator- DHI). Time-sag or push down effect is another observation that can be seen in the gas seepage area that leads to an image seem like a bowl shape structure. Application of seismic attributes such as Chaos, Variance-Coherence and Local Structural Dip Attribute are helpful in detecting unique responses caused by the gas presence, as well as improved the interpretation in the affected area. We present a summary for all 10 individual seepages interpreted in this work in Table 1 below. Results of this work can be very useful in early stages of seismic interpretation, be it for academic purposes or for field monitoring/development in the industry. In the near future, on surface analogue (outcrop) studies will be conducted to further understand the occurrence of gas seepage within the reservoir and we believed that a gas-seepage model for the clastic reservoir can be developed. This has been proven informative in the study of gas seepage as presented by [10].

Acknowledgement
We would like to thank Petronas Sdn. Bhd (Malaysian Petroleum Management) for providing the data for this work and Universiti Teknologi Petronas for the funding of this research. This is a preliminary work from a research grant 0153AA-D67, Investigation of Hydrocarbon Leakage through Seismic Data Analysis and On-Surface Field Analogue.

Table 1: Summary of interpreted gas seepages on the seismic data used in this study

Label	Approximate Size	Seismic Characteristics	Fault-Related ?	Remarks
Seepage-1	Wide: 0.5-10.0km Tall: up to 8.0km	Vertical Wipe-out Zone with Acoustic Blanking and Discontinuity Zone; Push-up effects at the edge & Push-down effects within the zone.	Possible fault/fractures related	Largest and most obvious gas seepage in the data. (more likely a cloud of gas) . Migrated Gas from deeper.
Seepage-2	Wide: 2.5-3.0km Tall: ~0.5-1.0km	Dim-Spot, Sand-Paper effects (chaotic), push-down effect	Yes	Related to Seepage-1 through Fault. Migrated Gas.
Seepage-3	Wide: 0.5-2.5km Tall: 0.5-1.0km	Vertical-Dim Zone, chaotic and sand-paper effects	Yes	Migrated from deeper gas seepage through fault
Seepage-4	Wide: 0.5-1.5km Tall: 2.0-3.0km	Bright Spot at the top of the leakage, Low-Continuity	Yes	Migrated from deeper gas seepage through fault
Seepage-5	Smaller seepages distributed laterally	Bright Spot at the top of the leakage, Sand-paper effects	No	Localized seepage within the sediments
Seepage-6	Wide:~2.0km Tall: 1.0km	Vertical dim zone, Low continuity, bright-spot at the top of the leakage, time-sag.	No	A crater with paleo-seabed depression
Seepage-7	Wide: ~1.3km Tall:2.0km	Vertical dim zone, Low continuity, bright-spot at the top of the leakage, time-sag.	Yes	A crater with paleo-seabed depression
Seepage-8	Wide:~1.5km Tall:~3.0km	Bright Spot at the top of the leakage, Sand-paper effects.	Yes	Migrated from deeper gas seepage through fault
Seepage-9	Smaller seepages distributed laterally	Laterally low-continuity zone, Lateral-dimmed zone with acoustic blanking and bright spot at the top of the leakage	No	Localized Gas seepage
Seepage-10	Smaller seepages distributed laterally	Lateral-dimmed zone with bright spot at the top of the leakage	No	Localized Gas seepage

References

[1] B. M. Schroot, G. T. Klaver, and R. T. E. Schüttenhelm, "Surface and subsurface expressions of gas seepage to the seabed—examples from the Southern North Sea," *Marine and Petroleum Geology,* vol. 22, pp. 499-515, 2005.

[2] A. Gartrell, Y. Zhang, M. Lisk, and D. Dewhurst, "Enhanced hydrocarbon leakage at fault intersections: an example from the Timor Sea, Northwest Shelf, Australia," *Journal of Geochemical Exploration,* vol. 78-79, pp. 361-365, 2003.

[3] D. Ghosh, M. F. Abdul Halim, M. Brewer, B. Viratani, and N. Darman. (2010) Geophysical Issues and Challenges in Malay and adjacent basins from E & P Perspective. *The Leading Edge*. 436-449.

[4] G. A. Logan, A. T. Jones, J. M. Kennard, G. J. Ryan, and N. Rollet, "Australian offshore natural hydrocarbon seepage studies, a review and re-evaluation," *Marine and Petroleum Geology,* vol. 27, pp. 26-45, 2010.

[5] A. L. Anderson and L. D. Hampton, "Acoustics of gas-bearing sediments I. Background," *The Journal of the Acoustical Society of America,* vol. 67, pp. 1865-1889, 1980.

[6] H. Løseth, M. Gading, and L. Wensaas, "Hydrocarbon leakage interpreted on seismic data," *Marine and Petroleum Geology,* vol. 26, pp. 1304-1319, 2009.

[7] M. Hovland and A. G. Judd, *Seabed pockmarks and seepages: impact on geology, biology and the marine environment*, 1988.

[8] M. Alarfaj and D. C. Lawton, "Interpreting Fault-Related Gas Leakage," 2014.

[9] A. H. Abdul Latiff, D. Ghosh, and S. F. Jamaludin, "Structural Enhancement in Shallow Gas Cloud Region " presented at the AeroEarth 2014, Bali, Indonesia, 2015.

[10] H. Løseth, L. Wensaas, B. Arntsen, N.-M. Hanken, C. Basire, and K. Graue, "1000 m long gas blow-out pipes," *Marine and Petroleum Geology,* vol. 28, pp. 1047-1060, 2011.

Improving Thin Bed Identification in Sarawak Basin Field using Short Time Fourier Transform Half Cepstrum (STFTHC) method

O. Nizarul[1], M. Hermana[1], Y. Bashir[1] and D.P. Ghosh[1]
Centre for Seismic Imaging (CSI), Department of Petroleum Geoscience, Universiti Teknologi PETRONAS, Tronoh, Perak, Malaysia.

E-mail: ni87ideas@gmail.com

Abstract: In delineating complex subsurface geological feature, broad band of frequencies are needed to unveil the often hidden features of hydrocarbon basin such as thin bedding. The ability to resolve thin geological horizon on seismic data is recognized to be a fundamental importance for hydrocarbon exploration, seismic interpretation and reserve prediction. For thin bedding, high frequency content is needed to enable tuning, which can be done by applying the band width extension technique. This paper shows an application of Short Time Fourier Transform Half Cepstrum (STFTHC) method, a frequency bandwidth expansion technique for non-stationary seismic signal in increasing the temporal resolution to uncover thin beds and improve characterization of the basin. A wedge model and synthetic seismic data is used to quantify the algorithm as well as real data from Sarawak basin were used to show the effectiveness of this method in enhancing the resolution.

1. Introduction

Stratigraphic interpretation of thin layers had not became attention until late 1970s when advances in field acquisition techniques rendered seismic amplitude interpretation more reliable. In case of our field study in Sarawak basin, specifically on clastic reservoir, deepwater exploration shows a good prospect. In Silantek Formation, the Basal Sandstone Member, Temudok Member and Upper Silantek Redbed Member showed an alternating thin beds of siltstone and sand [1]. The same trend can also be seen in Temodok Member [2].
Shell Sarawak by using Thomas-Stieber thin bed analysis increased the STOIIP estimate by 18% which was mostly contributed to by significant increment in net-to-gross (NTG) and improved estimations of porosity and saturation in one of the acquired block in Sarawak [3]. But imaging thin beds is an issue of resolution although the thin bed analysis has now become the routine in petrophysical analysis for Malaysian field [4]. A thin bed imaging gives interpreter brighter chance to produce a good petroleum model or in other cases predicts pitfalls that will hinder the petroleum system efficiency.

1.1 Thin Bed Definition

Seismic resolution is one of the key elements to estimate the quality of seismic data. Even with the advance technology of seismic acquisition through which we can obtain 50 Hz dominant frequency

[1] ni87ideas@gmail.com

seismic data, definition of 10 m beds are still a difficult task. Widess [5] considered a thin bed as one where the complex waveform across it does not differ significantly from the derivative of the convolving wavelet itself.

This definition is useful for thin bed detectability studies, but causes problems when it comes to thin bed resolvability considerations. At the bed thickness Widess [5] first considers a bed to become a "thin" bed, i.e., when the bed thickness is about l/8; the apparent thickness is actually l/4.6 which is the peak-to-trough time of the derivative of a Ricker wavelet [6]. Koefoed and de Voogd [7] refined thin bed definition and suggest that a bed is considered to be thin if the amplitude response, as a function of thickness, deviates less than 10% from a linear relationship.

A thin layer, formed by the intrusion or sedimentary process, commonly exhibit opposite reflectivity. If the separation of the layer boundaries is large enough, say exceeding one-eighth of the dominant wavelength of the propagating wavelet, and if it can be resolvable, the volume of the layer can be estimated by analyzing the configuration of the composite wavelets. Nevertheless, a thin layer of unresolvable thickness will very possible be ignored and become invisible on the seismic section [8]. Farr (1976) also emphasize that a bed as thin as l/40 may be detected. It should be fully understood, that the apparent thickness remains at l/4.6.

An important aspect of thin beds delineation is the frequency content of a composite wavelet made up of superimposed reflections from closely spaced interfaces [9]. Therefore, we apply STFTHC method to increase bandwidth of frequencies and later improved vertical resolution of the section for better interpretation work and seismic to well correlations.

2.2 *STFTHC technique formulation*

STFTHC algorithm was first proposed in Sajid [10] and was implemented in the Malay basin which also experience thin bed problem. The algorithm was made after an extensive comparative study on other well established signal enhancement algorithm comprising of Fanbeam transform [11], Wavelet transform [12], Ridgelet Transform [9] Curvelet transform [13] and Radon Transform [14]. The basis of STFTHC is combining the feature of Short Time Fourier Transform (STFT) with Half Cepstrum (HC) to to smooth and expand the frequency spectrum at each translation of the spectral decomposing window. The key idea is to replace the amplitude spectrum with its logarithm in each window of the STFT, the logarithm of the amplitude spectrum provides the opportunity to study the low energy and high frequencies information in frequency domain. The algorithm is the implementation of the Half Cepstrum (HC) algorithm on each slice of seismic trace. This is achieved by the merging of HC algorithm, originally from Cepstrum [15] with STFT algorithm [16]. Half Cepstrum (HC) is user independent frequency broadening algorithm which expands the frequency bandwidth by implementing logarithm on the amplitude spectrum of the trace.

2. Methodology

2.1 Method and Available Data

This frequency broadening technique depends upon the frequency contents in the seismic waveform. By incorporation of STFT, this logarithmic frequency broadening is implemented on each amplitude spectrum of the trace spectrogram. Reconstruction from this modified spectrogram leads to non-stationary high resolution seismic trace without greatly boost in white noise. The workflow on how we test the algorithm as seen in Figure 1. In the Half Cepstrum algorithm, frequency bandwidth is expanded in four simple steps as described below.

1. Fourier transform of the input signal,
2. Implementation of logarithm to expand bandwidth and smoothing,

3. Energy balancing with respect to the original amplitude spectrum. But before that the amplitude spectrum is brought to positive side,
4. Reconstruction back of the signal from broadening amplitude spectrum while using original phase spectrum. Below is the final algorithm of STFTHC:

$$STFTHC(t) = \iint_{-\infty}^{\infty} \left(HC_c(\tau, f)\gamma(t-\tau)e^{2\pi ft} df dt \right)$$

Where t = time, f = frequency in hertz, x(t) = time domain signal, h(t) = Gaussian Spectral decomposing window, τ = translation of Gaussian window along time axis, γ(t) = reconstruction window

The Gaussian windows of the windowing function:

$$W_{gw} = exp\left\{ -\left(\frac{n-b}{w}\right)^2 \right\}$$

where n = number of samples in a trace, b = position of the window, w = half width of the window

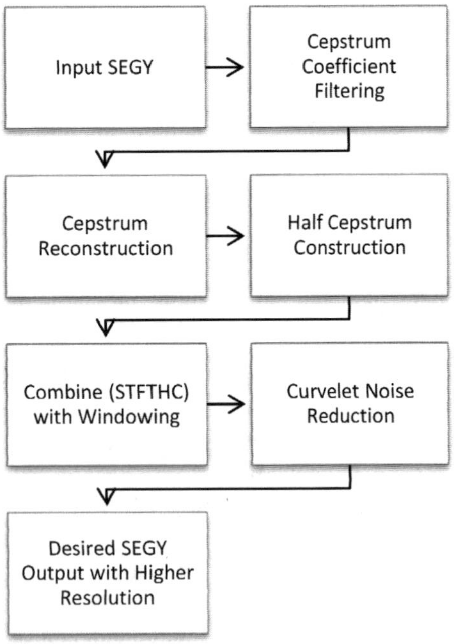

Figure 1: Flowchart of the working procedure we used to test the algorithm effectiveness.

In the STFTHC, the selection of the spectral decomposing window plays an important role and its size must be greater than the expected wavelet duration.

A time reflectivity method was used to create a synthetic seismic of known time thickness. It convolved with Ricker wavelet (zero phase) with the desired parameters (Figure 2a) and the resultant reflectivity (Figure 2b). The synthetic seismic created in Figure 3 is of 35 Hz dominant frequency with signal to noise ratio of 3 (Figure 2c). It contains 4 layers, layer 1 and 2 are the thin layer with same polarity while layer 3 and 4 produce the wedge model.

The produced synthetic model later was tested with STFTHC method to increase the resolution of the section by improving bandwidth.

For the real data testing, a section from Sarawak basin was used. Sarawak basin, as the adjacent Malay basin suffer thin beds problem, where thin coal beds are intercalated between reservoir sand shale layers and well data provides good control for our algorithm testing as most of the synthetic generated from well logs correlates well with the seismic. With the assumption that the proposed algorithm after bandwidth expansion will increase resolution of the layer. The data used is seen in Figure 6.

3. Result and Discussion

After the application of method we have test our data, the resolution was enhanced on both synthetic and real data. The synthetic section shows a boost in frequency content without generating gain for the noise shown below (Figure 3 and 4). In case of the synthetic section, which have S/N ratio = 3, the

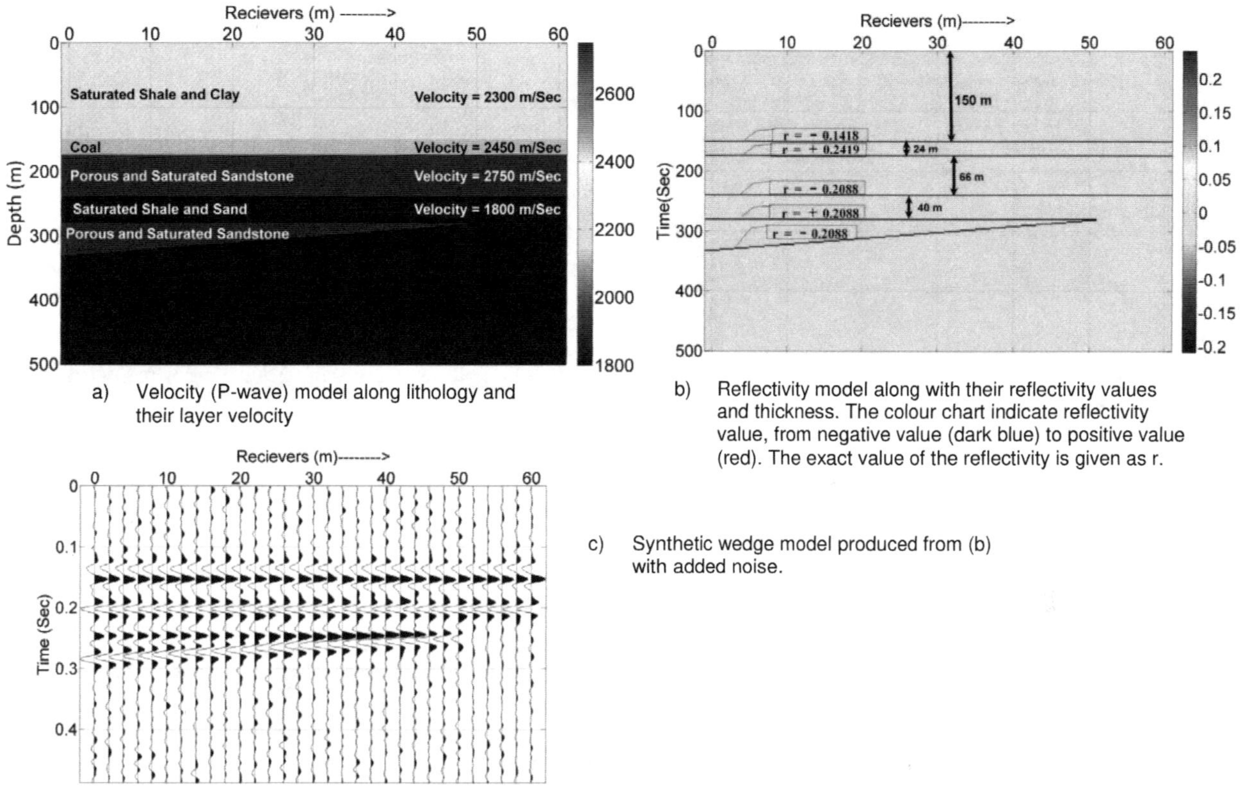

a) Velocity (P-wave) model along lithology and their layer velocity

b) Reflectivity model along with their reflectivity values and thickness. The colour chart indicate reflectivity value, from negative value (dark blue) to positive value (red). The exact value of the reflectivity is given as r.

c) Synthetic wedge model produced from (b) with added noise.

Figure 2: Velocity (P-wave) and Reflectivity model used to create synthetic wedge model in (c).

output of the test enhanced the visibility ranges from 60 Hz – 150 Hz, where thin beds usually fall in between the value.

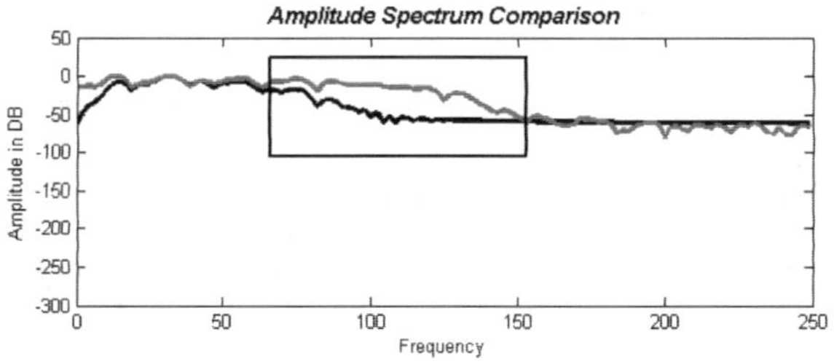

Figure 3: The amplitude spectrum comparison of original signal (black) and enhanced signal (red). Note the frequency ranges from 60 Hz to 150 Hz was boosted.

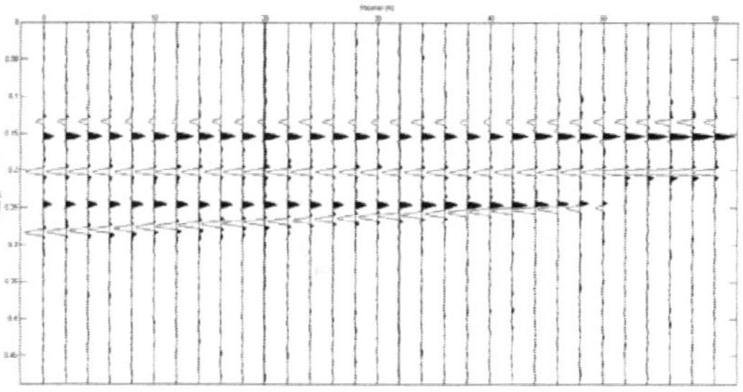

Figure 4: After STFTHC processing, the model now can have an enhanced resolution by gaining of the frequency content ranges from 60 Hz – 150 Hz, causing better resolution of thinner bed in the wedge (from black line to blue line)

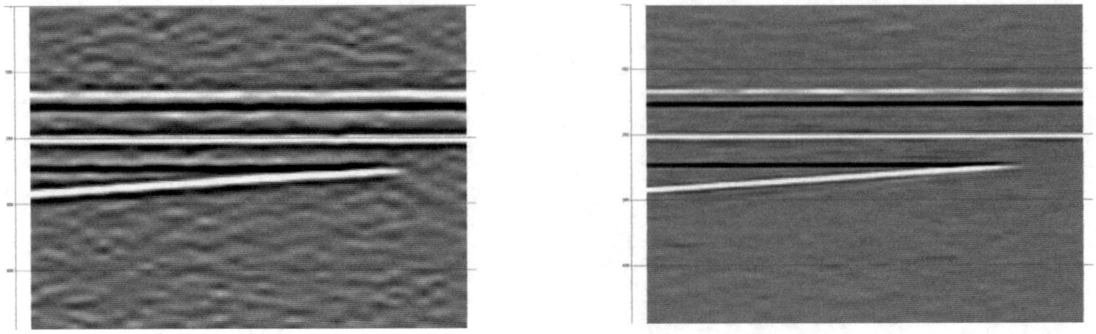

Figure 5: The enhanced resolution after application of STFTHC method (right) with clear reflectors for better subsurface interpretation of the wedge model.

In the real data for a seismic section in Sarawak, we expect to recover frequencies ranges from 10 Hz to 80 Hz (Figure 6). The range of frequency is very useful in thin bed tuning in seismic data processing and interpretation.

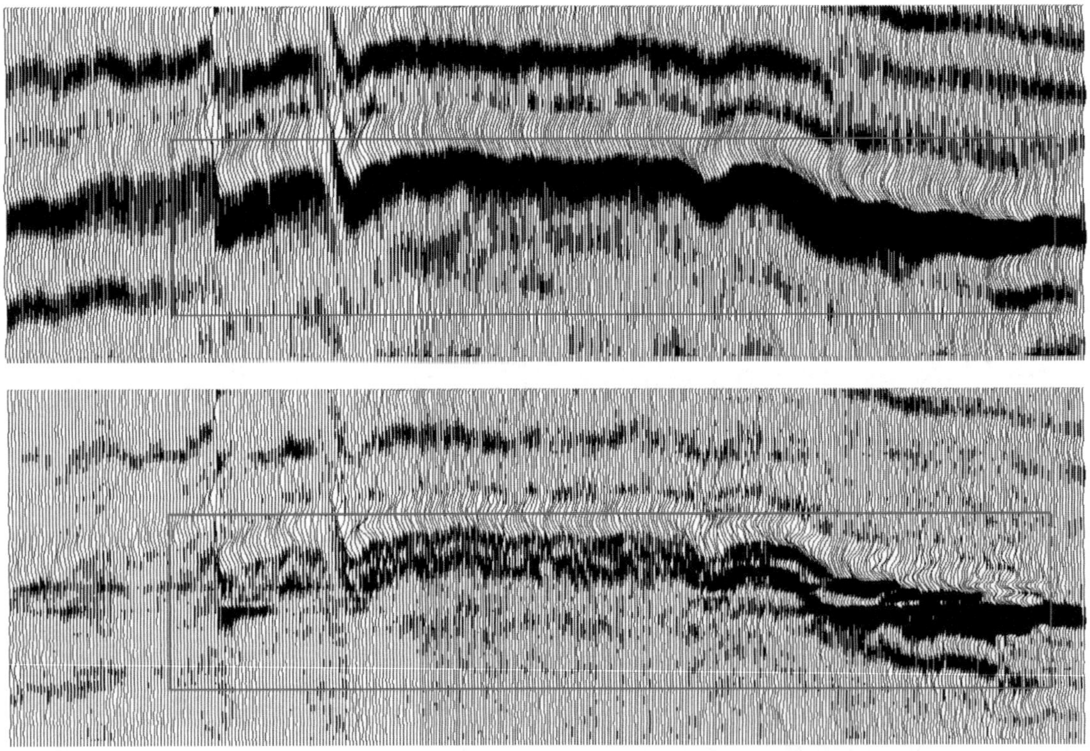

Figure 6: Real seismic section before STFTHC application (top) and after STFTHC application (below). The bed become more separable to unveil the thin bedding in between. Seen on the below right the bed become connected. With permission from PETRONAS.

4. Conclusions

We have demonstrated that the STFTHC method used help us in increasing the bandwidth of the seismic data for thin bed delineation. The resolution achieved through this algorithm depends upon the information preserved in the seismic waveform; therefore we can call them as data driven seismic resolution algorithms. This spectral enhancement method provide subsurface interpreter with more detailed data to uncover the once unresolved images of thin beds. STFTHC works in frequency domain, providing more flexibility for user to use it in spectral study with other domain for better and reliable interpretation.

Acknowledgments

We are thankful to Dr. Muhammad Sajid for his valuable suggestion on the algorithm, to PETRONAS for the permission on using the data and handful of shared knowledge on this subject matter, Universiti Teknologi PETRONAS (UTP) for giving permission to publish this work and the Centre for Seismic Imaging (CSI) which provide the facility for this research. In this work, we have used MatLab® and other tools developed by highly dedicated people from around the world like CREWES, WaveLab and SeisLab to enhance the speed of computing and make our work more efficient.

References

1. Tan, D.N., *The Lubok Antu Melange, Lupar Valley, West Sarawak: a Lower Tertiary subduction complex.* Bulletin of the Geological Society of Malaysia, 1982. **15**: p. 31-46.
2. Honza, E., J. John, and R. Banda, *An imbrication model for the Rajang accretionary complex in Sarawak, Borneo.* Journal of Asian Earth Sciences, 2000. **18**(6): p. 751-759.
3. Anijekwu, C., Carmalt, R., Lai, J. & Robail, F., *SF-30 Thin bed analysis - CaseStudy.* 2012.
4. Ghosh, D., et al., *Geophysical issues and challenges in Malay and adjacent basins from an E & P perspective.* The Leading Edge, 2010. **29**(4): p. 436-449.
5. Widess, M., *How thin is a thin bed?* Geophysics, 1973. **38**(6): p. 1176-1180.
6. Kallweit, R. and L. Wood, *The limits of resolution of zero-phase wavelets.* Geophysics, 1982. **47**(7): p. 1035-1046.
7. Koefoed, O. and N. De Voogd, *The linear properties of thin layers, with an application to synthetic seismograms over coal seams.* Geophysics, 1980. **45**(8): p. 1254-1268.
8. Chang, C., Shih R. & Chang, Y., *Characteristics of Reflectivity Strength on a Thin Bed.* TAO, 1996. **Volume 7**(Number 3): p. 269-276.
9. Chung, H. and D. Lawton, *Frequency characteristics of seismic reflections from thin beds.* Canadian Journal of Exploration Geophysics, 1995. **31**(1-2): p. 32-37.
10. Sajid, M. and D. Ghosh, *Logarithm of short-time Fourier transform for extending the seismic bandwidth.* Geophysical Prospecting, 2014. **62**(5): p. 1100-1110.
11. Zhuang, T., et al., *Fan-beam and cone-beam image reconstruction via filtering the backprojection image of differentiated projection data.* Physics in Medicine and Biology, 2004. **49**(24): p. 5489.
12. Daubechies, I., *The wavelet transform, time-frequency localization and signal analysis.* Information Theory, IEEE Transactions on, 1990. **36**(5): p. 961-1005.
13. Starck, J.-L., E.J. Candès, and D.L. Donoho, *The curvelet transform for image denoising.* Image Processing, IEEE Transactions on, 2002. **11**(6): p. 670-684.
14. Deans, S.R., *The Radon transform and some of its applications*. 2007: Courier Corporation.
15. Childers, D.G., D.P. Skinner, and R.C. Kemerait, *The cepstrum: A guide to processing.* Proceedings of the IEEE, 1977. **65**(10): p. 1428-1443.
16. Griffin, D. and J.S. Lim, *Signal estimation from modified short-time Fourier transform.* Acoustics, Speech and Signal Processing, IEEE Transactions on, 1984. **32**(2): p. 236-243.

Sedimentary Facies of the West Crocker Formation North Kota Kinabalu-Tuaran Area, Sabah, Malaysia

Azfar Mohamed[1], Abdul Hadi Abd Rahman[2], Mohd Suhaili Ismail[3]
Universiti Teknologi PETRONAS, 31750, Tronoh, Perak, Malaysia

E-mail: mohamed.azfar@yahoo.com, hadi_rahman@petronas.com.my, suhaili_ismail@petronas.com.my

Abstract. Newly outcrops exposed in the West Crocker Formation have led to the detail sedimentolgical analysis of the formation. Eight sedimentary facies have been recognised in which it was divided into three main groups: (1) sand-dominated facies (F1-F2), (2) poorly-sorted unit mixed sand and mud-dominated facies (F3), and (3) mud-dominated facies (F4-F5). These are: F1- graded sandstone (massive to planar laminated), F2-ripple-cross laminated, wavy and convolute lamination sandstone, F3-chaotic beds of mixed sandstone and mudstone blocks and clasts, F4-lenticular bedded of sandstone, and F5-shale. The studies of the formation has come out that it was deposited in a sand-rich submarine fan with specific location located at (1) inner fan channel-levee complex; (2) mid-fan channelised lobes, and (3) outer fan.

1. Introduction

Sabah which is located at the Northern part of Borneo Island lies within the active tectonic plates of Pacific, Philippine, Eurasia and Australia. The formation of the NW Borneo accretionary belt is closely related to the opening of the South China Sea. The West Crocker Formation of Oligocene to Early Miocene age which is mostly concentrated in the western part of Sabah is the main focus in this research. This sand-rich turbidite formation was deposited in an accretionary fore-deep basin forms part of this accretionary belt. This deep-water succession is well exposed over a large part of the coastal ranges of NW Sabah area can be up to 10 km in thickness (Stauffer, 1967). Newly exposed outcrops scattered at the North Kota Kinabalu to the Tuaran area, will be the focus in detail sedimentological analysis (Figure 1.0).

2. Geological Setting of the study area

The West Crocker Formation in NW Borneo had undergone two distinct periods of basin filling (Mazlan, 1999). It was first evolved during the Paleocene to Eocene period when deep-water sediments of the Rajang-Embaluh Group (Trusmadi Formation and East Crocker Formation) were deposited. These formed part of an accretionary complex due to the subduction of the South China Sea margin beneath Borneo at the North Borneo-Luzon Arc. The ongoing continuous subduction and compression processes caused the Rajang-Embaluh Group accretionary complex to be uplifted, folded and eroded which resulted in the formation of a major unconformity during the Late Eocene.

Due to ongoing subduction and compression, a second NE-SW trending accretionary complex was developed outboard of the topographic range represented by the newly uplifted Rajang-Embaluh Group. The complex comprised sediments of the West Crocker Formation (Sand-dominated facies) which is laterally equivalent with Temburong Formation (Mud-dominated facies) (Stauffer, 1967; Hutchison, 1989).

During Early Miocene, East Sabah was rifted and infilled with shallow marine sediments. At that time, the collision of Reed bank Terrane with Rajang Embaluh Group resulted in uplifting of the West Crocker Formation. The uplifting of Northwest Borneo occurred during Middle Miocene due to ongoing subduction of the South China margin. This resulted in the uplift, folding and thrusting of the

West Crocker Formation. The deformed, folded, imbricated and thrusted deep-water of the West Crocker Formation (Leong, 1999) which represents a northeast-southwest trending, elongate trough, part of a large submarine fan, built out northward from the Sunda Shelf in Eocene to Oligocene times, which was accreted against Borneo in the Late Oligocene to Early Miocene to form part of the Northwest Borneo Subduction Complex (Tongkul, 1987).

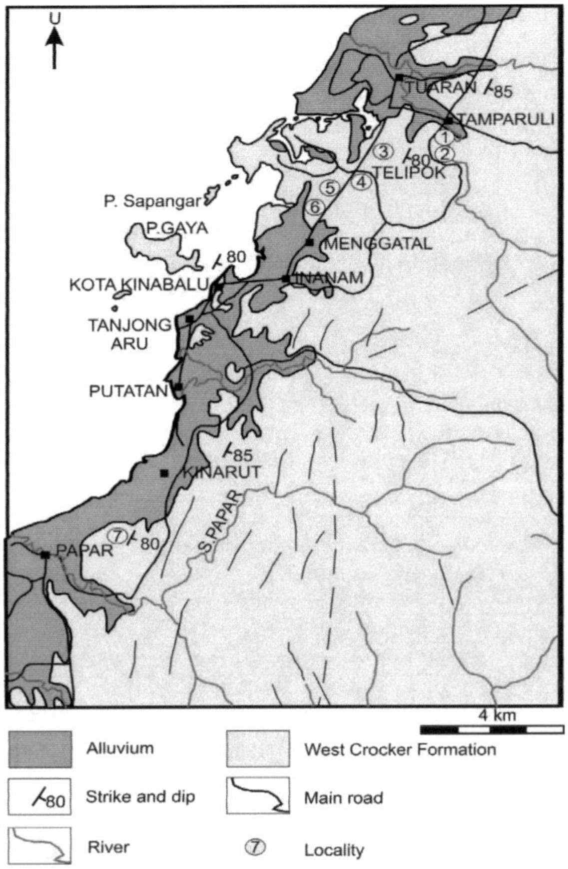

Figure 1.0: Geological mad and study area and outcrops locations. Outcrop location: 1-SMK Taman Ria, 2-Kg. Setinggan, 3-Jalan Telipok Lama, 4-Jalan Universiti Prima, 5-Jalan Sulaman, 6-KFC, 7-Lok Kawi.

3. Methodology

The geological mapping involved details descriptions, measurements and sampling at seven exposed localities near the town of Kota Kinabalu and Tuaran area, NW Sabah. The sections with total measured thickness of 227.25 m were extensively been logged, and based on 1:50 and 1:100 scales. The seven selected localities are: 1) SMK Taman Ria (WCF-1), 2) Kampung Setinggan (WCF-2), 3) Jalan Telipok Lama (WCF-3), 4) Jalan Universiti Prima (WCF-4), 5) Jalan Sulaman (WCF-5), 6) KFC (WCF-6),7) Lok Kawi (WCF-7).

The sedimentology logging was documented with emphasis on the lithology, sandstone thickness, grain size, morphology of the beds contact, sedimentray structures, bioturbation style and internal characterictics. These features will be used to interpret facies and facies association of the succesions and the depositional will be construcuted for the West Crocker Formation, NW Sabah, Borneo.

4. Sedimentary facies
The West Crocker Formation can be divided into eight sedimentary facies.

4.1. Graded Sandstone (Massive to Parallel Laminated)

This facies consists of light grey, fine to coarse grained with moderate sorted of sandstone. It shows medium to thick-bedded with individual thicknesses between 3 to 5 meters (Figure 2A). Basal contacts of beds are sharp, or locally erosive with abundant of granule to pebbles which can result the normal grading (Figure 2B). Generally, these beds are massive or display as feint horizontal stratification (Figure 2C). Soft sediment deformation such as dewatering structures, dish structures, dewatering pipes and mudclasts are common either at the middle or at the top of beds. The sandstones show parallel lamination, current-ripple and climbing current ripple lamination developed in the finer-grained at the upper part of this beds (Figure 2D). The flute cast and groove marks are common. Massive character implies of rapid deposition from highly concentrated turbulent flow (Tongkul, 1987). The poor development of internal deposition structures indicates that these sandstones were deposited rapidly by processes of traction and fallout from short-term turbidite flow (Collenette, 1958; Tongkul, 1987).

4.2. Ripple-Cross Laminated, Wavy and Convolute Lamination of Sandstone
F3 contains of light grey, very fine-grained sandstone with interbedded with hemipelagic mudstones. This facies resembles alternation of laminated mudstone with sharp, planar bases with lack of flute casts or tool marks. This facies is typically 0.5 to 1 m in thickness. Internally, F3 consists of intervals of current-rippled to climbing current ripple very fine-grained sandstone which repetitively alternate with wavy and convolute mud drapes of flaser bedding (Figure 3E).

This F2 can be interpreted as lower density turbidity current deposit with extensive uni-directional flows in respond to sediment supply and hemipelagic setting. Current-ripples to climbing-current ripples indicates that late stage uni-directional traction modification processes occurred during deposition of F2 suggesting that sediment supply to the bed was relatively high with some additional energy and balanced the transport of sediment across the bed which allowing ripples to aggrade (Middleton, 1993). The presence of wavy mud drapes indicate the more extensive tractional modification with more incoming sediments input and these uni-directional flow were pulsatory in nature and allowing mud-sized particles to settle out or fallout of suspension on a form of mud drapes (Lowe, 1982).

4.3. Chaotic Bed of Mixed Sandstone-Mudstone Blocks and Clasts
This facies refers to a chaotic combination of sandstone and mudstone. Mudstone commonly appears in dark grey-coloured while sandstone is yellowish or light-grey in colour. Fine-grained sandstone significantly contains more mud than high-density turbidite sandstone. The contacts may be gradational or sheared. This facies consists of clast rich units, with 1 to 2 m in thickness. Internally, this facies is always chaotic and poorly sorted, structureless, and ungraded (Figure 3F). The sandstone and mudstone clast shows tabular to sub-spherical. The coal form as finely-disseminated chips within matrix exists in this facies. This facies is sharply overlain by sandy high-density turbidites of facies 1 sandstone unit.

Facies 3 is interpreted as a debrite, which was deposited by a flow either in laminar or weakly turbulent. The ungraded nature of both of the structures and clasts in different size suggested that this facies is deposited as a single phase of the flow in which whole mass undergoes large and continuous deformation by freezing off en masse, rather than by progressive settling (Muddler and Alexander, 2001). The variety clast shapes within this facies and their internal structure information is strong evidence to show that this facies is formed of debris flow in which normally tend to be supported by matrix strength, dispersive pressure, and buoyancy (Middleton, 1993). Clasts in this facies are internally deformed suggesting that the clasts are poorly lithified at the beginning of the deposition and later it may be deformed by internal shearing or the interaction of clasts within the flow.

4.4. Lenticular Bedded Sandstone

This sandstone lenses is yellowish in colour while the shale is dark grey. These occurrences are generally fine- to medium-grained sandstone. The composite thickness of this facies ranges from 3 to 5 m. The sandstone lenses are about 0.5 to 2 cm and shale layer is about 4 to 8 cm (Figure 3G). The unit extends laterally about 5 m. Most of sandstone beds have sharp basal and top contacts and are capped with shale layers. Internally, the sandstone beds consist of parallel lamination and climbing lamination. The deposition of the alternations between sandstone and shale requires fluctuating flow conditions, where it represents a quiet and low energy of turbidity current, which has fall-out and repetitively. The variable thickness in this facies suggested that it formed at a moderately strong, intermediate cycle of flow. The existence of parallel lamination and climbing ripples in the sandstone beds imply a waning turbidity current.

4.5. Shale

F5 is shale in dark grey-coloured. The F5 unit range from 0.5 to 1 m in thickness. Its displays as a sharp top and bottom contact of the beds. It lateral continuity is about 5 m. Stratigraphically, F8 commonly occure above F1 or F2. F5 is interpreted to present deposition of suspension fallout of finer grained particles. In turbidite system, deposition of coarse sediments may leave a residual suspension of fine grained sediments. This residual fine grained sediment will wane and tend to deposit under lower energy of turbidity currents. This fine grained material could also associate with and transported within the flow with response to a rapid settle down (Amy *et al.*, 2005).

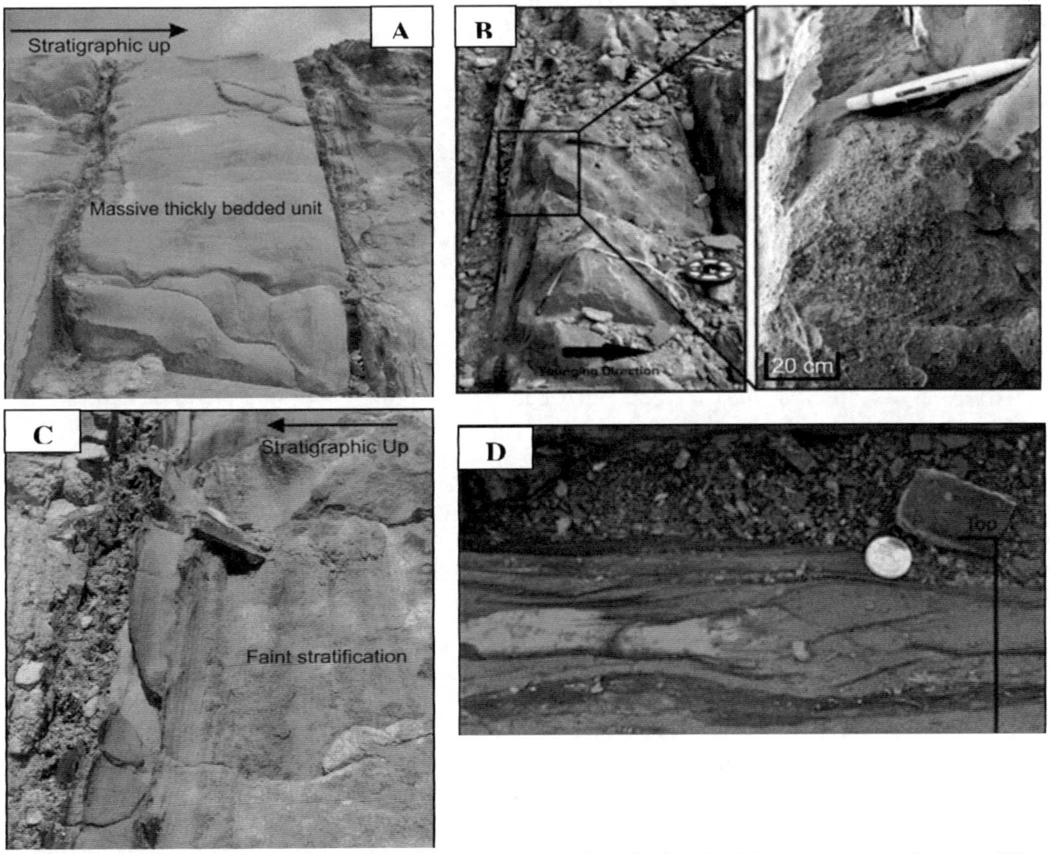

Figure 2: Main characteristics of facies 1: (A) massive thickly bedded unit of sandstone; (B)massive sandstone with granule to pebbles lags; (C) medium to fine-grained sandstone with feint parallel stratification; (D) ripple cross-stratification, climbing ripples in the sandstone unit.

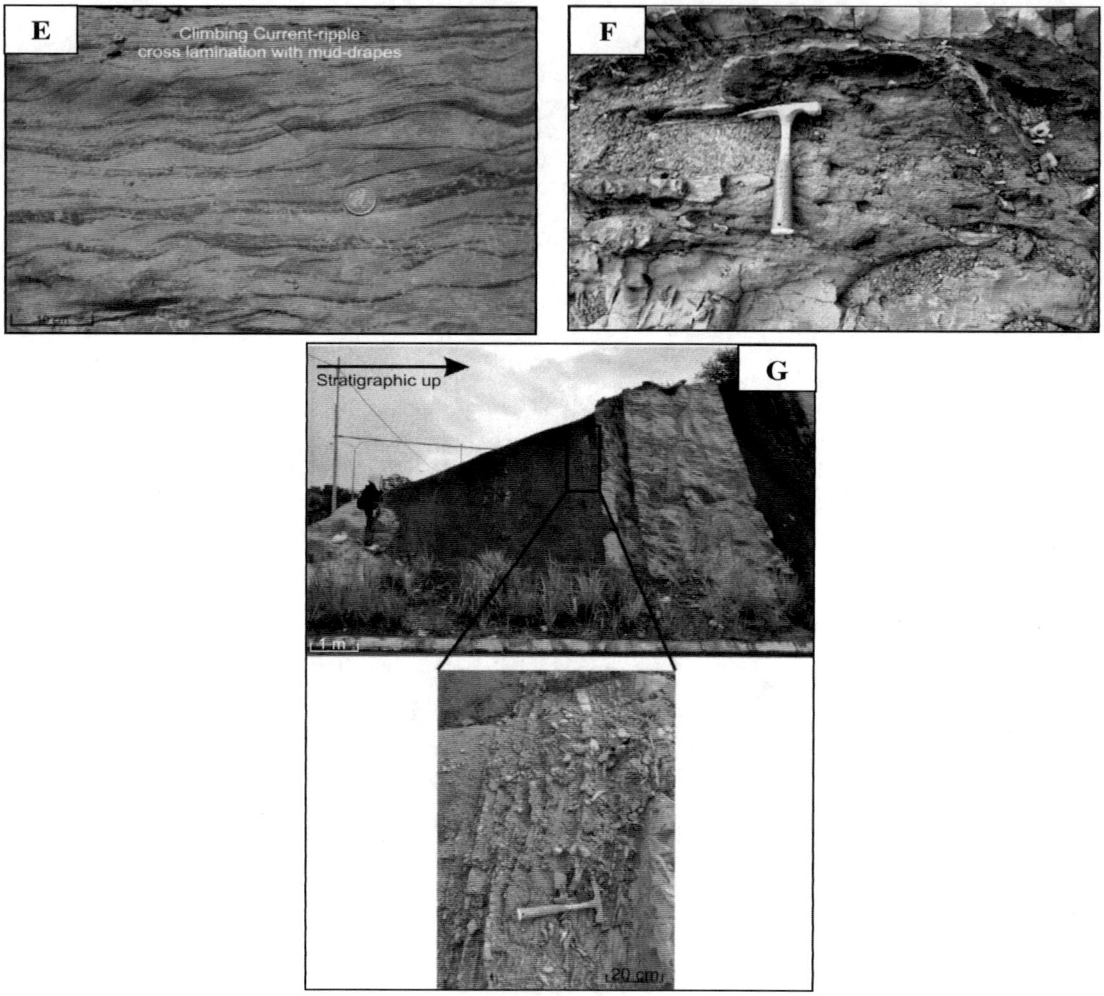

Figure 3: (E) F2 shows climbing current ripple interbedded with thin layer of mud; (F) F3 chaotic unit; (G) F4 mud-dominated lenticular bedded facies.

5. Depositional Environment

The West Crocker Formation, NW Sabah is characterised by high amount of sandstone content dominated by thick amalgamated, channel-levee, channelized and distal lobe sandstones. Based on the documented five sedimentary facies, a sand-rich submarine fan depositional model is proposed for the West Crocker Formation, as following (Mutti and Normark, 1987) model. This model is generally divided into Inner, Mid, and Outer fan (Walker, 1978).

6. Conclusions

1. Five facies were identified based on the lithology, sandstone bed thickness, grain size, bed contact, sedimentary structures and bioturbation style. These are: 1) F1- Graded sandstone (Massive to Parallel Laminated), 2) F2- Ripple-cross laminated, wavy and convolute lamination sandstone, 3) F3- Chaotic beds of mixed sandstone and mudstone blocks and clasts, 4) F4- Lenticular bedded sandstone, 5) F5- Shale.

2. Based on the integration studies, the studied sedimentary successions were deposited in a sand-rich submarine fan specifically located at (1) inner fan channel-levee complex; (2) mid-fan channelised lobes, and (3) outer fan distal lobes.

7. Acknowledgement
I wish to thank PETRONAS and Universiti Teknologi PETRONAS for research funding.

8. References

[1] Stauffer, P.H., 1967. Studies in the Crocker Formation, Sabah. Geological Survey Borneo Region Bulletin, 8, 1-13.
[2] Mazlan M., Leong K.M, Anuar A., 1999. Sabah Basin. In: The Petroleum Geology and Resources of Malaysia. Petroliam Nasional Berhad (PETRONAS), Malaysia, 499-543. [3] Sze S M 1969 *Physics of Semiconductor Devices* (New York: Wiley–Interscience)
[3] Hutchison, C.S., 1989. "Geological Evolution of South-East Aisa". Oxford monographys on geology & geophysics, No 13, Clarendon Press, Oxford. [Reprinted in softback by Geological Society of Malaysia, 1996].
[4] Leong, K.M., 1999. Geological Setting of Sabah. In: The Petroleum Geology and Resources of Malaysia. Petroliam Nasional Berhad (PETRONAS), pp. 475- 497.
[5] Tongkul. F., 1987. The sedimentological and structure of the West Crocker Formation in the Kota Kinabalu area, Sabah. In: GEOSEA IV Proceedings, Jakarta 1987, Indon. Assoc. Geol., 135-156.
[6] Tongkul, F., 1991. Tectonic evolution of Sabah, Malaysia: Journal of Southeast Asian earth Science, v.6, p.395-405, doi: 10.1016/0743-9547(91)90084-B.
[7] Collenette, P., 1958. The geology and mineral of the Jesselton-Kinabalu area, North Borneo, British Borneo Geological Survey memoir 6.
[8] Lowe, D.R. 1982. Sedimentary Gravity Flows: II. Depositional models with special reference to the deposits of high-density turbidity currents. Journal of Sedimentary Research, 52: 279-297.
[9] Middleton, G.V., 1993. Sediment deposition from turbidity currents. Annual Review of Earth and Planetary Sciences 21, 89–114.
[10] Mulder, T. and Alexander, J. (2001). The physical character of subaqueous sedimentary density flows and their deposits. Sedimentology, 48, 269-299.
[11] Amy, L.A., Talling, P.J., Peakall, J., Wynn, R.B., Arzola Thynne, R.G., 2005. Bed geometry used to test recognition criteria of turbidites and (sandy) debrites. Sedimentary Geology 179, 163–174.
[12] Mutti, E., and W. R. Normark, 1987, Comparing examples of modern and ancient turbidite systems: Problems and concepts, in J. K. Leggett and G. G. Zuffa, eds., Marine clastic sedimentology: Concepts and case studies: London, Graham and Trotman. p. 1–38.
[13] Walker R.G., 1978. Deep water sandstone facies and ancient submarine fans: models for exploration for stratigraphic traps. AAPG Bull., 62, 932-966.

IOP Conf. Series: Earth and Environmental Science **30** (2016) 012005 doi:10.1088/1755-1315/30/1/012005

Tectonic development of the Northwest Bonaparte Basin, Australia by using Digital Elevation Model (DEM)

Ali Wahid[1], Dr. Ahmed Mohamed Ahmed Salim[2], Dr. Gamal Ragab Gaafar[3], AP Wan Ismail Wan Yusoff[4]

[1,2,4] Universiti Teknologi PETRONAS, Seri Iskandar, 32610 Tronoh, Perak Malaysia,
[3] Petronas Carigali Exploration, Kuala Lumpur, Malaysia.

E-mail: ali.wahid@live.com,

Abstract. The Bonaparte Basin consist of majorly offshore part is situated at Australia's NW continental margin, covers an area of approx. 270,000km^{2}. Bonaparte Basin having a number of sub-basins and platform areas of Paleozoic and Mesozoic is structurally complex. This research established the geologic and geomorphologic studies using Digital Elevation Model (DEM) as a substitute approach in morphostructural analysis to unravel the geological complexities. Although DEMs have been in practice since 1990s, they still have not become common tool for mapping studies. The research work comprised of regional structural analysis with the help of integrated elevation data, satellite imageries, available open topograhic images and internal geological maps with interpreted seismic. The structural maps of the study area have been geo-referenced which further overlaid onto SRTM data and satellite images for combined interpretation which facilitate to attain Digital Elevation Model of the study area. The methodology adopts is to evaluate and redefine development of geodynamic processes involved in formation of Bonaparte Basin. The main objectives is to establish the geological histories by using digital elevation model. The research work will be useful to incorporate different tectonic events occurred at different Geological times in a digital elevation model. The integrated tectonic analysis of different digital data sets benefitted substantially from combining them into a common digital database. Whereas, the visualization software facilitates the overlay and combined interpretation of different data sets which is helpful to reveal hidden information not obvious or accessible otherwise for regional analysis.

Introduction

The Bonaparte Basin is the northernmost sedimentary basin situated at North West Shelf of Australia [1] having up to 15 km of Phanerozoic sediments [2] [3]. A period of Middle Miocene folding, incorporating extensional wrenching and related with the collision of the Australian and Asian continental plates, affected most of the Northwest Shelf. During this period, an important heat pulse occurred along the Northwest Shelf [4]. The major structural elements of Bonaparte Basin comprised

[1] ali.wahid@live.com

of Plover Shelf, Darwin Shelf/ Bathrust Terrace, Petrel sub-basin, Malita Graben, Sahul Platform, Sahul syncline (Figure. 1). The Plover Shelf is the easterly extension of the Kimberley Block and contains a thin cover of Cretaceous and Tertiary sediments. A thin wedge of Permian to Recent sediment onlaps onto the Darwin Shelf and Bathurst Terrace [5]. The Petrel Sub-basin is bounded shorewards by major basin margin faults, and extends northwestwards into the Sahul Syncline. The sub-basin contains mainly Late Carboniferous to Triassic sediments; since the Jurassic, the depocentre of the Bonaparte Basin has moved northwestward. The Petrel Sub-basin tilts regionally to the northwest [6]. A thick Jurassic sequence occurs within the Malita Graben; it was initiated in the Late Triassic and culminated in the Callovian. Block faulting has produced a NE-trending horst and graben topography, with drape in the overlying Late Cretaceous and Tertiary sediments. The Sahul Platform is an uplifted area of faulted Triassic/Permian sediments, overlain by a Late Jurassic to Tertiary sequence [6]. A large hiatus occurs in the Late Triassic due to erosion, which resulted from uplift from the end of the Triassic to Middle Jurassic. The platform has formed an essentially stable block since the Early Jurassic. The Sahul Platform is separated from an equivalent platform to the west (Ashmore) by the Sahul Syncline. Permo/Triassic sediments thin onto both these platforms, while the Sahul Syncline contains a thick sequence and represents an extension of the Petrel Sub-basin (Figure 3). This outer region was uplifted during the Triassic to Middle Jurassic but foundered from the Middle Cretaceous. when westerly prograding carbonate shelf deposition became predominant [7].

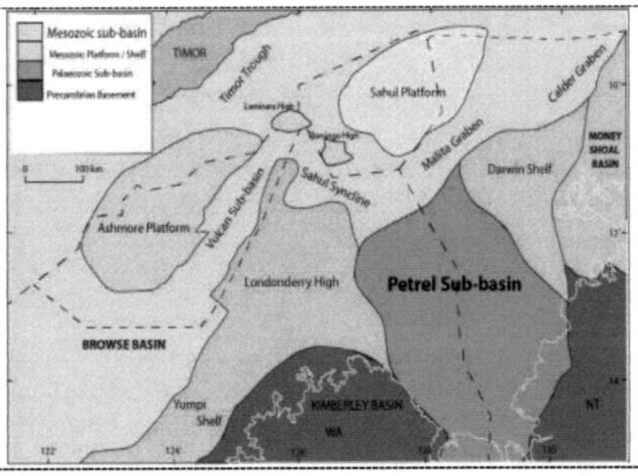

Figure 1. Structural elements of the Bonaparte Basin

Tectonic elements of Cenozoic mainly contain broadly NE–SW structural axes and trending faults of Miocene to Recent age [5]. This most recent deformation phase formed as a consequence of the partial subduction of the Australian Plate with the Banda Arc and northward collision [8]. Though, the timing of the collision events between the Australian margin and the Banda Arc region is still discussable, as the Cenozoic period was geologically complex and climatic evolution in SE Asia [8]. According to [3] the Middle–Miocene (12 Ma) onwards is the time of the subduction of the northward- dipping Australian continental margin. The present work intended to illustrate stages of deformation in the Northern Bonaparte Basin, in order to estimate their effect on the geometry of the continental shelf and the sediment distribution from the onset. Palaeozoic and Mesozoic structural elements can thus be considered as potential first order control elements on the recent evolution of the northern Bonaparte Basin, for instance, the Sahul Platform and Darwin Shelf interbasinal highs, and the NE–SW-trending Malita Graben low (Figure 3).

2. Methodology and Materials/Data

The surface information consists of the integrated satellite and topographic images were used for regional structural analysis in the NW Bonaparte basin. The data has been incorporated in the mapping of tectonic events. Whereas, the available Landsat TM/ETM images with 1.5 arc second resolution were used for overview surface mapping and higher resolution images (Spot data) were used locally. Shuttle Radar Topographic Mission (SRTM) digital data, with a 2500m ground resolution, were worn for geomorphological analysis as well as mapping and identification of faults and structures. Satellite images blended with SRTM data to facilitate interpretation. As, the SRTM data are particularly useful for geomorphologic and structural analysis. Available open-file and internal geological and tectonic maps have been geo-referenced and were overlaid on SRTM data and satellite images for combined interpretation. The above data have been Geo-referenced and loaded into Global Mapper. Combined information from satellite imageries and digital elevation data as well as 3D views significantly facilitated in geological understanding by revealing details and information not available otherwise. Structural geological interpretations were compared to integrate different geological events occurred in different time.

Geological maps was digitally arranged and cross sections was generated which is based on up and low anomalies of secondary structures, geological structural maps and information of previous geological surficial. The interpretation facilitate the consideration on the geological settings in terms of tectonics and geodynamic of the Bonaparte Basin.

3. Results and Discussions

The northern limit of the Bonaparte Basin is marked by the present continental shelf margin immediately south of the Timor Trough. Subdivisions of the Bonaparte Basin have been integrated and incorporated in the digital elevation model with the help of satellite images and the maps which was georeferenced, in the study area. The area spans the transition from a passive to a collisional margin as the result of northward migration of the Indo-Australian plate that occurred since the Late Oligocene [3]. Both the timing of events of this Neogene collision [3] and the distribution of structural styles that resulted in the Bonaparte basin [2] have been previously investigated and are still highly debated.

3. 1. Geological Features of Bonaparte Basin

During the Cambrian, the deposition starts in the onshore portion of the Petrel Sub-basin because of the widespread subsidence and volcanism and subsidence in the region. The development of Bonaparte intially starts as an intra-cratonic rift or sag basin in the Cambrian to Ordovician, which can be seen onshore and in the intermediate offshore basin (Figure 2). There are different key events including extension, volcanism, subsidence etc. occurred throughout the evolution of the Bonaparte Basin.

From Devonian to Carboniferous the northwest-trending Petrel Sub-basin was formed due to surface extension. The imprint of the older trend was because of the extension with northeast structural grain. This is the development time Malita Graben and proto-Vulcan Sub-basin (Figure 3).

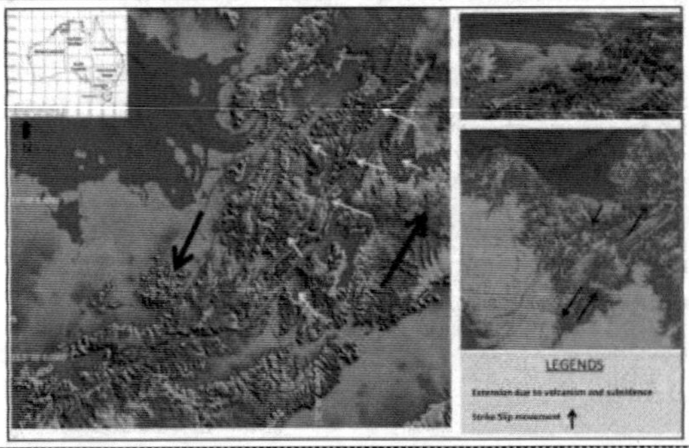

Figure 2. Onshore rift events in the NW Bonaparte Basin

The highly faulted and structured sequence is formed because of initial sedimentary fill from Late Devonian to Early Carboniferes, which is confined between the original bounding faults of the rift. With the help of available geological maps, the Late Devonian- Early Carboniferous extensional faults have been georeferenced on the study area.

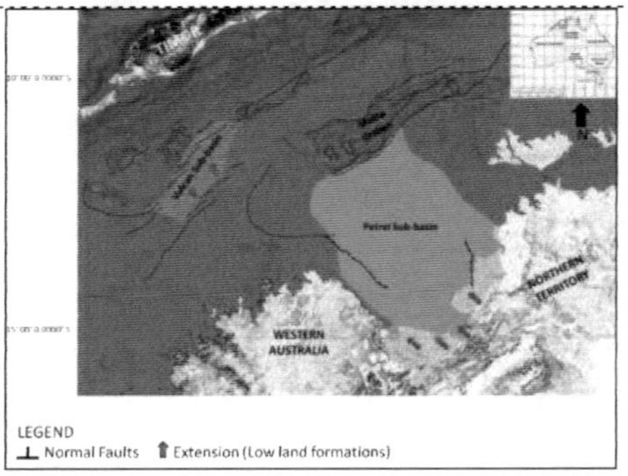

Figure 3. Late Devonian to Early Carboniferous extension responsible of formation of Petrel sub-basin alongwith Malita Graben and Vulcan Sub-basin

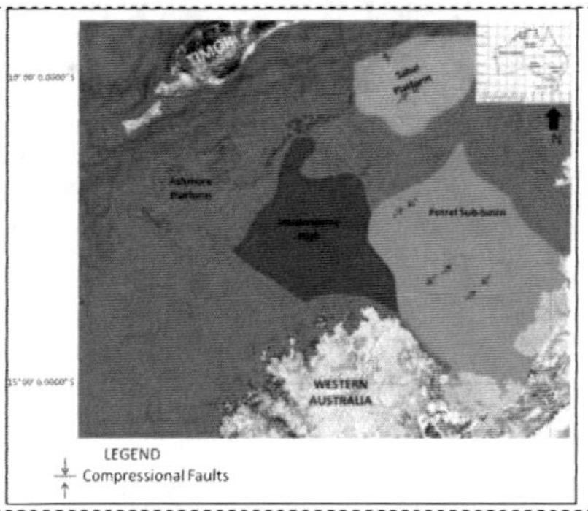

Figure 4. Late Triassic compression in Bonaparte basin responsible for uplifting of Ashmore, Londonderry High and Sahul platforms, and the margins of the Petrel Sub-basin

Further, a compressional event in the Late Triassic occured which results the uplift and erosion of the Londonderry High, the Ashmore and Sahul platforms, and the southern margins of the Petrel Sub-basin. The main rifting and extension of upper-crust (Late Devonian-Early Carboniferous), bacame a time of sag-dominated deposition (Carboniferous to Permain), punctuated by a series of renewed upper and/or lower crustal extension-sag cycles. The integration of post-rift thermal cool-down, sediment loading, probable movement on basement faults and crustal thinning was driven by sag-dominated phase. During the Permian age the crustal extension ceased, which is responsible for thermal subsidence phase sedimentation (until the Late Traissic). A major compressional reactivation event affected the Petrel Sub-basin during the Late Triassic to Early Jurassic, reactivating faults along the basin margins (mainly in the southwest).The sea level starts droping duirng compressional event until another extensional period starts at Late Jurassic [5]. This Mesozoic extension is responsible for major Jurrasic depocentres which includes Malita Graben, Vulcan Sub-basin, Sahul Syncline, Calder Graben and Malita Graben. During Early Cretaceous the thermal subsidence in the onset is repsonsible for fine grained thick wedge of clastic and carbonate sediments progradation from Cretaceous to Cenozoic across the offshore Bonaparte Basin [5].

As Late Cretaceous is the time of high sea level which is responsible for the carbonate deposition with clastic sedimentation. The clastic sedmentatin is because of the thermal subsidence occured in the onshore part of the basin. The compressional event responsible for the some of the Jurassic depocentre are overprinted in the model (Figure 4).

In the mid-Miocene the Australian continental plate collided with the Indonesian plate along the northern continental boundary, resulting in ENE-WSW compression [8].The strongly faulted northern margin of Sahul Platform and Timor Trough was formed during the last major events occured in the study area of Miocene age in which regional compression associated with the collision of the Southeast Asian microplates and Australia-India plate (Figure 5).

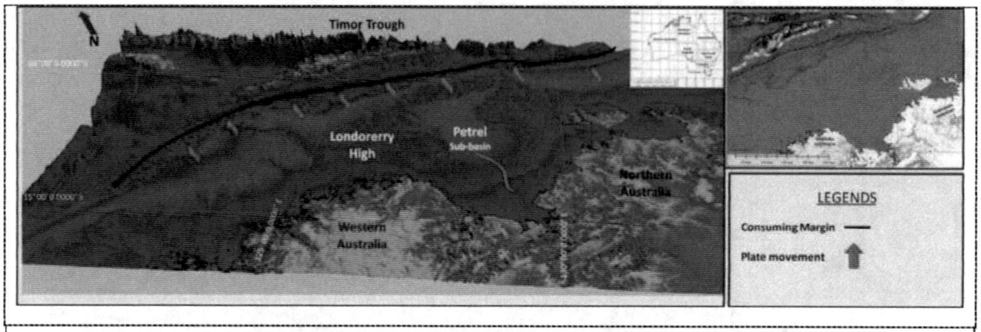

Figure 5. Regional compressional event related with the Australia-India and Southeast Asia microplates collision and formation of Timor Trough

3.2. Bonaparte Basin Evolution

All the tectonic events occured in the study area have been finally integrated in a Digital Elevation Model which includes the subbasins and plateforms of the Bonaparte Basin incorporated with the tectonic elements, occured at different geologic times (Figure 6).

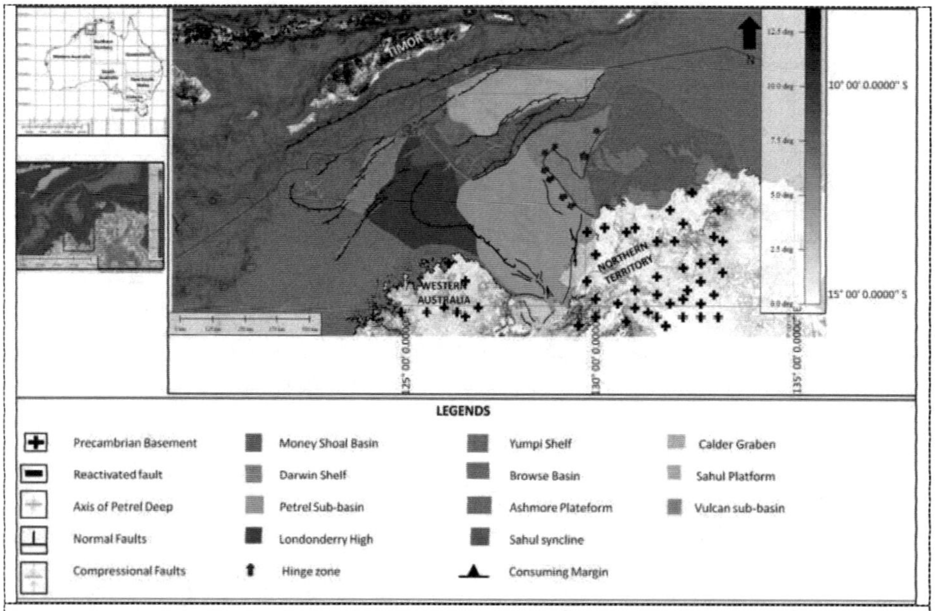

Figure 6. Tectonic Digital Elevation Model of Bonaparte Basin

Two E-W geological cross sections (A-A' and B-B') combined with Digital elevation model across NW Bonaparte Basin for the illustration of tectonic effect towards the surface have been established (Figure. 7). Left lateral shear between the Indo-Australian and Pacific Plates in the New Guinea region has resulted anticlockwise rotation of Australia and Northwest Shelf due to dextral shear imposition (right-lateral) is the cause of major structural development and fault reactivations (Figure. 7.a). During Post Early Jurassic, the petrel sub-basin mainly underwent slow sag. So, the anticlinal structure may have been created in Petrel-subbasin by inversion. Although it also is possible the structures may have formed as a result of salt movement and progressive salt withdrawal (Figure 7.b).

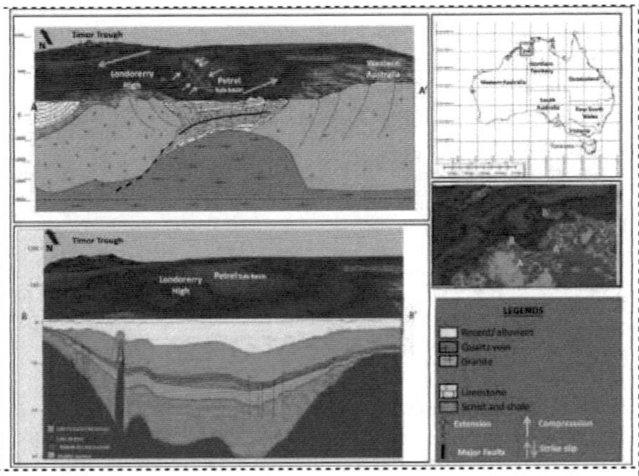

Figure 7. E-W geological cross sections across NW Bonaparte Basin

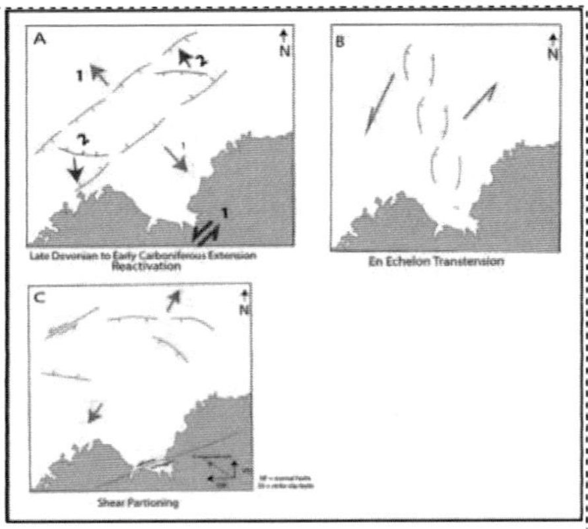

Figure 8. Extension events originate the several normal faults.

4. Conclusions

The study is helpful to build a Digital Elevation Model of Bonaparte Basin. The geometry of the system bounded by the faults which are close to a large pull-apart and recommended a strike-slip component. A series of major en-echelon folds, particularly at Northern Carnarvon Basin have developed as a result of the imposition of a regime of right lateral shear. Three possible hypotheses as shown in figure 8 suggested their presence, i) strike-slip faults are responsible for the reactivation of normal faults; ii) a dip-slip and strike-slip components are involved in the transtensional event consisting of faults; or iii) the oblique extensional stress contributed the shear partitioning into normal faults and strike-slip faults. Among the supposition, the reason of initial extensional event is because of shear partitioning is more favorable, because of strike-slip and dip-slip faults existence in the area.

5. Acknowledgement

The authors would like to acknowledge Universiti Teknologi PETRONAS for facilitation of this research and Department of Mines and Petroleum, Western Australia for providing data.

References

1. Longley, I., et al. The North West Shelf of Australia–a Woodside perspective. in The Sedimentary Basins of Western Australia 3: Proceedings of the Petroleum Exploration Society of Australia Symposium, Perth. 2002.
2. Shuster, M., et al., Neogene tectonics, greater Timor Sea, offshore Australia: implications for trap risk. The APPEA Journal, 1998. **38**(1): p. 351-379.
3. Hall, R., Cenozoic geological and plate tectonic evolution of SE Asia and the SW Pacific: computer-based reconstructions, model and animations. Journal of Asian Earth Sciences, 2002. **20**(4): p. 353-431.
4. Bultitude, R., Flood basalts of probable early Cambrian age in northern Australia. Volcanism in Australasia, 1976: p. 1-20.
5. Frankowicz, E. and K. McClay, Extensional fault segmentation and linkages, Bonaparte Basin, outer North west shelf, Australia. AAPG bulletin, 2010. **94**(7): p. 977-1010.
6. S.J.Cadman and P.R.Temple, Australian Petroleum Accumulations Petroleum Greenhouse Gas and Advice Group, 2003. **Report 5**(2nd edition).
7. Cadman, S. and P. Temple, TERRITORY OF ASHMORE & CARTIER ISLANDS ADJACENT AREA (AC) JOINT PETROLEUM DEVELOPMENT AREA (JPDA). 2004.
8. Keep, M. and D.W. Haig, Deformation and exhumation in Timor: Distinct stages of a young orogeny. Tectonophysics, 2010. **483**(1): p. 93-111.

Post-Stack Seismic Data Enhancement of Thrust-Belt Area, Sabah Basin

A.H. Abdul Latiff[1], S.N.F. Jamaludin[2], M.N.A. Zakariah[3]

[1, 2, 3] Centre for Seismic Imaging, Geosciences Department, Universiti Teknologi PETRONAS, Malaysia
Email: abdulhalim.alatiff@petronas.com.my

Abstract: In this paper, an integrated post-stack seismic data processing and interpretation for a complex thrust-belt area was proposed. The sequence was suggested due to poor seismic data quality of the Sabah basin area that was obtained after a pre-stack data processing sequences. This basin consists of a complex geological setting such as thrust-belt with steep dip reflector which is the main features of the region. In this paper, we outlined several methods used in the seismic data processing and interpretation such as amplitude recovery and frequency filtering for enhancing seismic data quality, and relative acoustic impedance, structural smoothing and wavelet coherency were used for attribute analysis. The outcome from this research aims at illuminating the hidden structures such as proper beds termination and faults systems that was heavily affected by low signal-to-noise ratio.

Keywords: Post-stack Processing, Attribute Analysis, Thrust-Belt Region

1. Introduction

In simple geological structure, the final seismic image obtained is generally good, with identification of rock properties, hydrocarbon traps and fault systems are relatively straightforward. The seismic data acquisition deployed for this area can be considerably cheap, while the usage of a standard seismic data processing sequence is sufficient for producing good images. With the era of easy oil is over, more seismic exploration are focusing on complex structure, deep and limited accessibility area as well as marginal oil and gas fields, to arrest the decline in hydrocarbon production. However, there are many hydrocarbon-producing fields in offshore Malaysia suffered from poor seismic data quality due to the presence of near surface anomaly. For example, several fields located in the Malay Basin are located underneath the shallow gas cloud anomaly and need additional processing and imaging steps in order to enhance the quality of the dataset [1].

In the offshore Sabah area, where it is located at the northern part of Borneo (**Figure 1**), is generally known as a complex geological area, which also producing a poor seismic data quality. Geographically, the Sabah Basin is bounded by West Baram Line (west) and Balabac Strait Fault (east), while extends north westwards beyond continental shelf, into the Sabah Through. Based on the study in [2], the tectonic evolution of Sabah Basin has pass through three major deformations along NW-SE and N-S compression axes, involving changes in stratigraphy, structural geology, paleo-magnetic and its depositional history. This basin is predominantly Middle Miocene sedimentary basin that evolve since

[1] Corresponding author

Late Cretaceous [3]. However the basement of the basin in deeper area is a relatively un-metamorphosed succession of turbidites, that belonging to the Oligocene-Lower Miocene West Crocker, Temburong and Kudat formations, which represent Crocker Fold-Thrust Belt.

The area of study is basically located in Central offshore Sabah, 150km from Kota Kinabalu, Sabah shoreline. It is situated adjacent to Inboard Belt to the west of the field of study, and consists of a tight, complex faulted asymmetric anticline structure along WNW-ESE trending with deep-seated thrust fault system. It was believe that these faults system as a result of strong basement induced wrench movements during the Late Miocene. The presence of steep dips along the fault planes caused the poor illumination of the seismic data. In addition, the prospects greater than 500m water depth are structurally complex, mostly due to toe-thrust plays with turbidite reservoirs. In previous years, several major discoveries have been made, with the fields such Gumusut-Kakap, Malikai, Ubah and Petai were discovered by Shell while Kikeh field was discovered by Murphy, which contributes approximately 15% of Malaysia's domestic production [4].

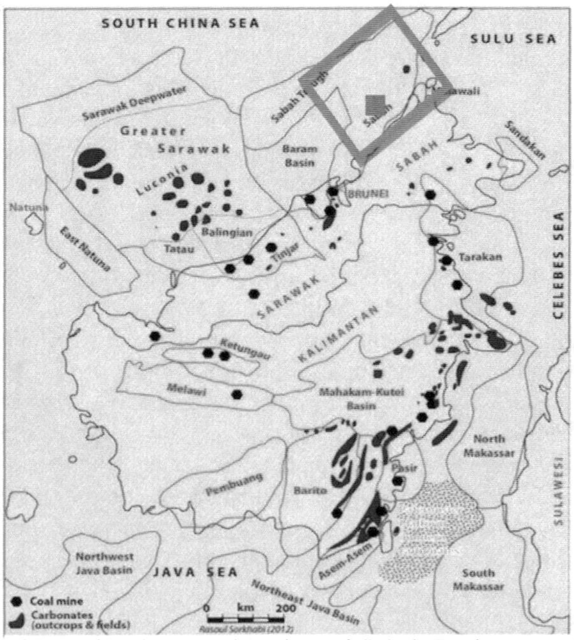

Figure 1: The map location of Sabah Basin (green rectangle), is located on the major segment of Sarawak-Brunei-Sabah continental margin. The field of study is indicated by orange box.

2. Imaging Issues

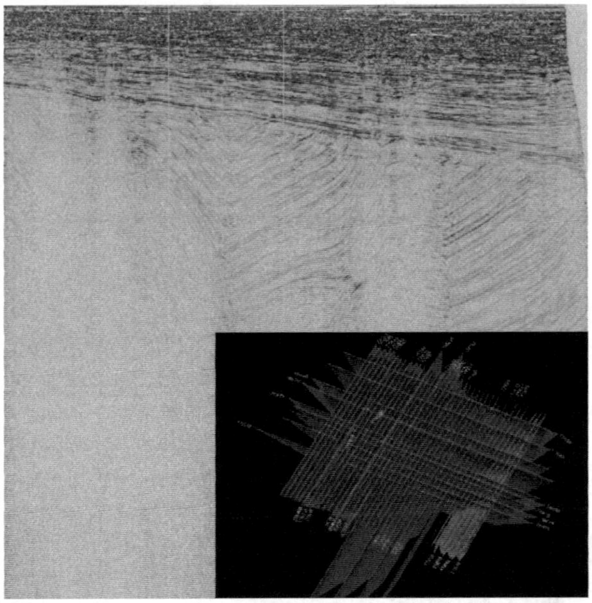

Figure 2: A 2D line of raw migrated seismic section of the study area. Noted that poor seismic illumination in the green box area. (Inset) The survey area comprising a set of 2D lines for the area of study in Sabah Basin. (Data permission from PETRONAS).

Since the area of study is regarded as a highly complex geology in terms of structure and stratigraphy [5], the existing 2D seismic data which was acquired in 1970's and 1980's, unable to address the uncertainties surrounding the fault zones due to poor signal to noise ratio. Consequently, it will raise some difficulties in interpretation, proposing well location and hydrocarbon volumetric calculation and estimation in the later exploration stages. Generally the vintages data in this area have a good image quality over the flanks of the structure, but exhibit a considerably poor image quality at the steeply dipping reflectors where the faulted area are situated (**Figure 2**). This is because, the steeply dipping beds along with complex faulting caused a rapid lateral velocity variations over the whole subsurface structure in this area.

Although several re-processing seismic data had been carried out to improve the data quality, the presence of almost 30^0 to 70^0 steep dips at deeper depth caused the wave propagation path to be reflected away from the projected survey area. For example, if the dipping reflector is located some 1000m beneath the surface level, the reflected wave will be detected at the receiver located 2500m away from the source. However, if the dipping reflector is situated 2000m beneath the sea level, the reflected wave can only be detected at a distance around 6km away from the source location. Since the seismic data of this field was acquired by using streamer length ranging from 2400m to 3000m, the reflected signal from these dipping reflector was not properly processed and imaged.

Post-stack seismic processing and pre-stack seismic processing are two different methods. The seismic data processing for the former method can only be conducted if all the seismic data had been sorted and stacked according to Common Depth Point-Offset sequence. In pre-stack seismic processing, the data enhancement sequences are to be carried out before the sorting and stacking process. In order to enhance the quality of subsurface imaging in the post-stack sequences, several signal processing methods were tested, before a new post-stack seismic data enhancement sequence was developed. The sequence which consists of several steps resembles the pre-stack processes such amplitude gain using programmable gain control and frequency filtering, were incorporated to improve the resultant data for better interpretation, particularly in the complex faulting zones that is highly ambiguous to interpret using the vintages dataset. In addition, three seismic attributes methods were implemented (relative acoustic impedance, structural smoothing and coherence-variance), for better structural anomaly delineation. The proposed post-stack processing steps for this particular complex area are as follow:

1. Amplitude-time gain to boost up the low amplitude energy at the deeper section of seismic data.
2. Butterworth band pass filtering in order to remove the low frequency noise that affecting some major reflectors as well as to remove high frequency anomaly noise signal.
3. Incorporating relative acoustic impedance that yield important information about lateral changes in lithology and porosity.
4. Apply structural smoothing to preserve important discontinuities such as faults or channels.
5. Apply seismic coherence attributes based on variance method. The coherence attribute will give an easier tracking and interpretation of faults within a 2D window.

3. Results and Analysis

3.1 Amplitude Recovery and Frequency Filtering

In the original seismic data which was obtained after a pre-stack time migration process (**Figure 2**), the seismic data was displayed according to its true amplitude reflectivity using variable density colour scheme. This caused low signal energy detected at certain depth as it decays rapidly once the seismic wave penetrates the earth and reflected back to the receivers. Thus, if the seismic wave reflected from the deeper reflectors were displayed with the same amplitude scale as that from the shallow reflectors, peaks and troughs from deeper events would be invisible to the eye. Therefore a programmable gain recovery method is required to normalize all amplitudes within a trace by looking at a time windows over that trace and adjusting the amplitude of events within the window relative to a chosen standard. This will boosted up the seismic signal according to the depth of the signal, i.e. deeper section will experience a greater amplitude gain compared to shallower section. In **Figure 3a**, the seismic signal beneath the unconformity was gain up accordingly, and a 'V" shape dipping reflectors were spotted. However, as being highlighted in **Figure 2**, the area within the green box, still experiencing poor illumination even though several energy boosting procedure were applied. This probably due to the presence of shale diapir which cause amplitude attenuation to the seismic wave travelled pass through the area. To prevent any artefacts that may cause from unnecessary gain function, it was suggested to apply a moderate gain function that enable only important steep dips to be interpreted.

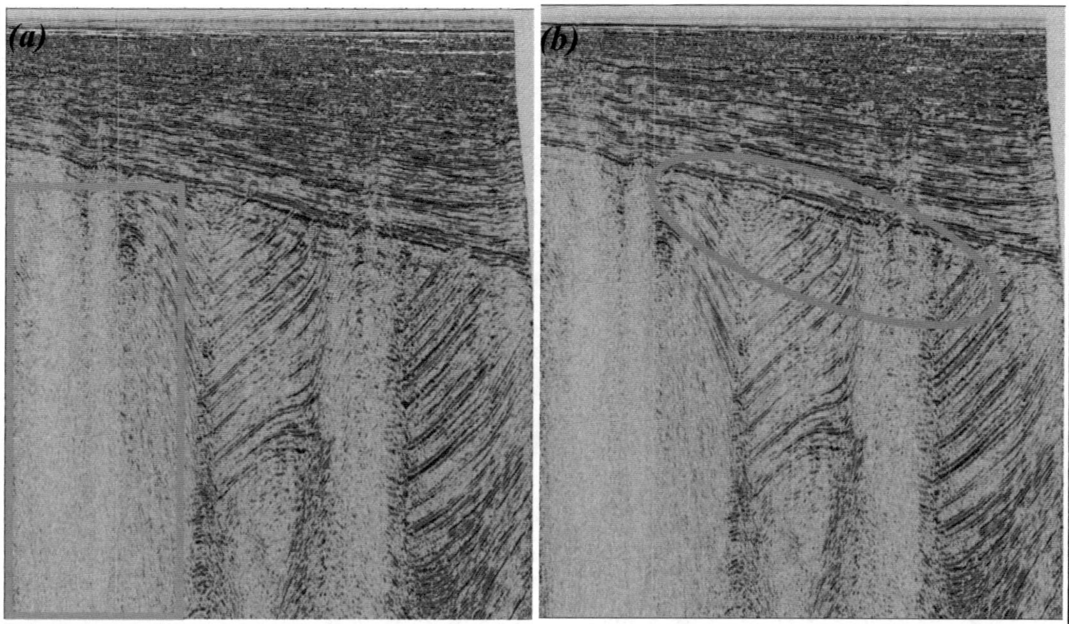

Figure 3: *(a)* The seismic section after amplitude recovery process has enable a steep dip reflectors to be imaged. However, some area still experiencing poor illumination (green box), probably due to shale diapir that caused loss of amplitude. *(b)* The seismic section after frequency filtering process which remove low frequency noises that masking the data (orange oval) as well as high frequency noises that produces artefacts on the data. (Data permission from PETRONAS).

The filtering procedure then was applied to the data in order to remove low frequency noises that has been masking the data as well as the high frequency anomaly that was created during pre-stack processing sequences. Basically, filtering in seismic data processing is a process of limiting the frequency content of the original signal. This process was carried out in order to differentiate the reflection signal of interest from the noises, multiples and artefacts, which lead to higher signal-to-noise ratio. Since the signal response recorded will be differ from field to field, each geophysicist must choose the correct frequency band required to highlights the area's seismic target. Thus, the highly subjective decision is totally depend on several testing parameters, such as the cut-off frequency limit, filter slope and the types of filter, i.e. band-pass, low-pass, high-pass or notch filter, before choosing the parameters that works for that particular field. Throughout the process, we keep in mind that noise content may also vary with time and position along the line, thus multiple filtering processes of varying time / distance, is sometimes needed. In field located in Sabah basin with complex geological setting, a 10 Hz low frequency cut-off, and 70 Hz high frequency cut-off was implemented along with 18 dB/octave and 27 dB/octave of low and high filter slope respectively. The outcome from this filtering, shown in **Figure 3b**, indicates the low frequency noises were remove from the seismic data (orange oval), thus give a better quality seismic image.

3.2 Relative Acoustic Impedance and Structural Smoothing

In the next data enhancement sequence, a standard seismic attribute in relative acoustic impedance was applied to the field dataset. The acoustic impedance which was produced after inverting the post-stack seismic amplitude, is a physical property expressed as a multiplication of medium density and acoustic wave velocity [6]. The acoustic impedance contrast between adjacent mediums controls the angle and amplitude of acoustic reflection and refraction at the considered interface. This inverted impedance section yielded some useful information about lateral changes in lithology and porosity. In **Figure 4a**, several seismic characteristic enhancement were seen, among them are reflectors which were correlated as a result of relative acoustic impedance implementation on the seismic section, particularly the angular unconformity reflectors (highlighted in orange box). In addition, several bright spots can be seen (in green circle), indicate a high reflectivity sediments compare to its surrounding, which also indicate that a potential hydrocarbon trap might existed in the area.

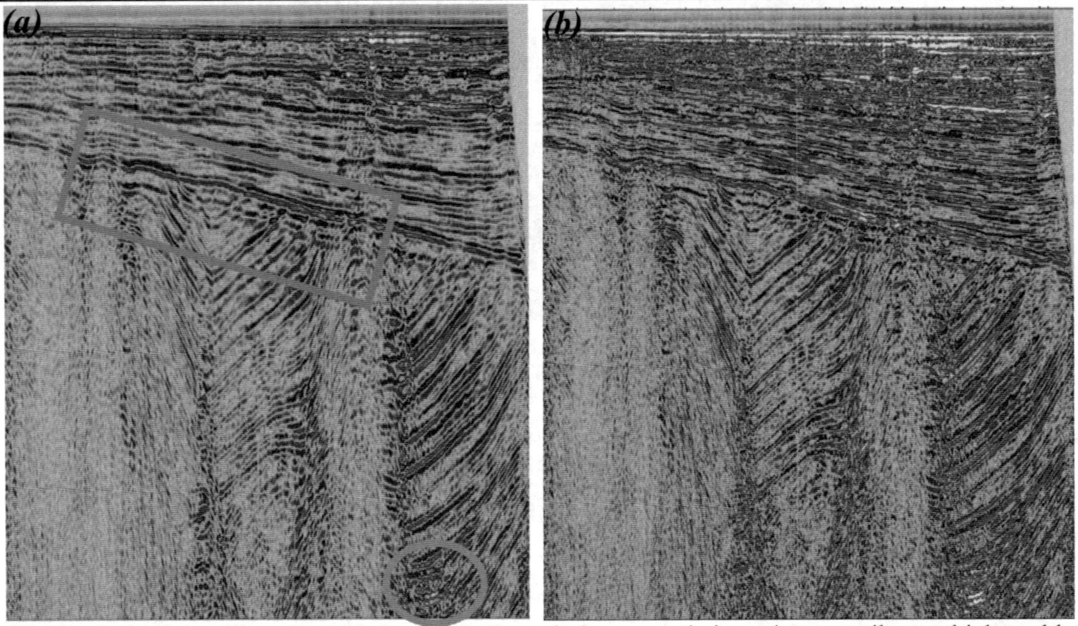

Figure 4: *(a)* Seismic section after incorporating relative acoustic impedance attribute which enable a better correlation of reflectors, particularly the unconformity layer over the shale diaper (orange box). *(b)* Then the structural smoothing was applied to the seismic section in order to preserve important discontinuities such as faults or channels, before final attribute analysis, seismic coherence was incorporated. (Data permission from PETRONAS).

Before the seismic coherence attribute was applied, a structural smoothing need to be applied to the whole seismic section. This method was chosen as it will reduce random noise while preserving structure without prior information of the structure orientation is needed. Principally, it measures the orientations of these beds and smooth the resistivity measurements along directions which lie along the measured orientations. Consequently, the boundaries between adjacent beds are preserved while noise streaks are removed. Hence, automatic detection techniques can be employed to identify beds or fractures and dip angles of the beds. The resultant structural smoothing for field data from Sabah basin (**Figure 4b**) shows a clearer reflections image, as more noises were removed, thus revealing a continuous horizon above the steep dip reflectors.

3.3 Seismic Coherence – Variance

Seismic coherence attribute is a measure of similarity between waveform or traces which is the response obtained from the seismic wavelet convolve with geology of the subsurface. A low coherency data indicate an abrupt changes in waveform which in turn represent faults and fractures system of the sediments and consequently will provide interpretation insights. Most of the time, coherence attribute was implemented on the seismic data with the target to enhance and impose a greater understanding of fault interpretation particularly in the complex geological area. From the result in **Figure 5**, it can be seen that a fault structure (indicate in orange arrows) actually existed in the shallow sedimentary layers (which is not visible in **Figure 2**). This prove that the fault line can be clearly interpreted when a proper seismic data processes were applied before the application of any volumes and structural attribute methods.

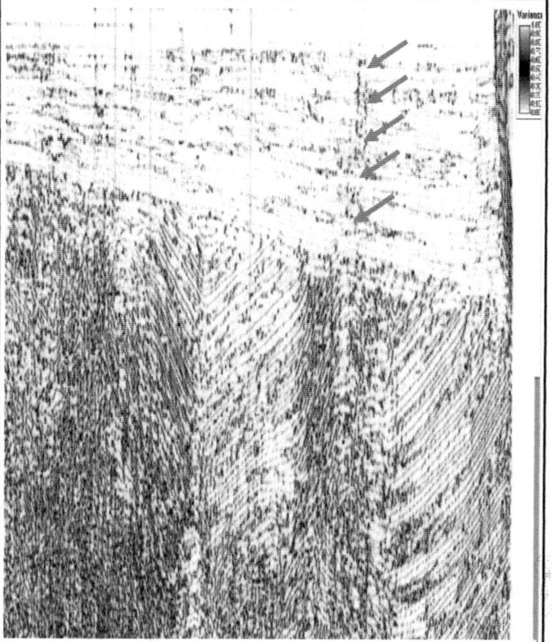

Figure 5: Figure shows the resultant seismic coherence attribute applied on the post-stack seismic data. The attribute implementation able to distinguish a fault system (along the orange arrows) that previously was invisible. (Data permission from PETRONAS).

4. Conclusion

Throughout this paper, we have demonstrate the benefit of post-stack seismic data processing implementation towards improving vintage seismic data. It is also proven as an effective methodology before any seismic attributes analysis on structural characterization was conducted. Without the implementation of noise and artefacts removal processes to the seismic data, the interpretation and fault detection process in the complex geological setting is extremely difficult as no obvious fault line can be recognized. Therefore the outcomes from this integrated processing and attribute analysis will provide a better data quality compared to separate data analysis and implementation. However, it is also suggested that a new 3D seismic data should be acquired to significantly improve the seismic data quality.

Acknowledgement

We would like to thank PETRONAS's Malaysian Petroleum Management (MPM) for providing the seismic data for this work and Universiti Teknologi PETRONAS for the funding of this research. This is a preliminary work from a research grant 0153AA-D70, Seismic Interferometry Analysis for Characterizing Principal Geological Units of Malaysia.

References

[1] A. H. Abdul Latiff, D. P. Ghosh, and S. N. F. Jamaludin, "Structural Enhancement in Shallow Gas Cloud Region," *IOP Conference Series: Earth and Environmental Science,* vol. 23, p. 012005, 2015.

[2] F. Tongkul, "Tectonic evolution of Sabah, Malaysia," *Journal of Southeast Asian Earth Sciences,* vol. 6, pp. 395-405, 1991.

[3] M. Madon, *The petroleum geology and resources of Malaysia*: PETRONAS, 1999.

[4] D. P. Ghosh, M. F. Abdul Halim, M. Brewer, B. Viratno, and N. Darman, "Geophysical issues and challenges in Malay and adjacent basins from an E&P perspective," *The Leading Edge,* vol. 29, pp. 436-449 2010.

[5] F. Tongkul, "The geology of Northern Sabah, Malaysia: its relationship to the opening of the South China Sea Basin," *Tectonophysics,* vol. 235, pp. 131-147, 1994.

[6] S. Chopra and K. J. Marfurt, *Seismic Attributes for Prospect Identification and Reservoir Characterization*: Society of Exploration Geophysicists, 2007.

IOP Conf. Series: Earth and Environmental Science **30** (2016) 012007 doi:10.1088/1755-1315/30/1/012007

A comparison on initial-value ray tracing and fast marching eikonal solver for VTI traveltime computing

S.Y. Moussavi Alashloo[1,*], D.P. Ghosh[1], Y. Bashir[1] and W.I. Wan Yusoff [1]

[1] Department of Petroleum Geoscience, Universiti Teknologi PETRONAS, Sri Iskandar, Malaysia

E-mail: y.alashloo@gmail.com

Abstract. The Earth's subsurface is an anisotropic medium where the velocity of seismic waves alters in different propagation angles. Omitting anisotropy in seismic imaging not only brings mis-positioning of migrated dipping events but also fails to retain dipping energy during dip-moveout. To account for the efficacy of seismic anisotropy in imaging, an anisotropic wave equation must be engaged. Seismic traveltime computing is fundamental of both Kirchhoff migration and tomography algorithms. Two main categories of traveltime computing involve traditional ray tracing methods and finite difference eikonal solvers. In this study, we present two techniques of initial-value ray tracing and fast marching eikonal solver in isotropic and vertical transverse isotropy (VTI) media, and a comparison between results is demonstrated for more evaluation. Although the ray tracing approach is able to compute multiple arrivals with great precision, the eikonal solver is faster and more robust for traveltime computation. Since the ray tracing result is not a deterministic solution and it depends on the initial circumstance, employing the eikonal solver method are more preferred and suggested.

1. Introduction

Hydrocarbon and geothermal reservoirs, and overlying strata are comprised anisotropic rocks. Considering anisotropy into account is necessary not only to avoid distortions in imaging, but also provides important information about lithology and fracture networks. One of the causes of anisotropy in sedimentary rocks is the thin layering. If the structure has horizontal layers, it is a TI medium with vertical axis of symmetry (VTI). One of the challenging problems in seismic imaging is the calculation of time taken by seismic wave for traveling from source to receiver. In P-wave velocity analysis and imaging, several traveltime computing algorithms have been developed for VTI condition [1-3]. There are two principal classes of traveltime computing which include ray tracing and eikonal solver methods.

The ray technique was firstly employed to study the propagation of the high-frequency elastic waves by Babich [4]. The main algorithms in ray tracing are defined as shooting and bending methods. The shooting approach utilizes a given initial direction and position for the ray and an interpolation tool to reach a certain point which is also known as initial-value method [5]. For modeling and imaging of complex structures by using shooting method, one can compute the multiple arrivals containing the most energetic wave which does not necessarily correlate to the first arrival. Sun and Schuster [6] employed the initial-value approach for the wave path migration. Julian [7] developed an initial-value method to track rays in inhomogeneous media which is applied by Engdahl [8] for earthquake analysis.

An efficient alternative to compute traveltimes is solving the eikonal equation by employing finite differences [9, 10]. Although eikonal solvers only provide the first arrival times, they can compute traveltimes between two points. An attempt to extend eikonal solvers for obtaining multiple arrivals are conducted by Bevc [11]. Various techniques are introduced to acquire the solution of the eikonal equation comprising single-pass methods, embedding methods, sweeping methods and iterative methods [9]. The main difference of these techniques is that how they cope with the complication of multi-valued solutions, finding the solution in the vicinity of cusps and discontinuities [12]. Anisotropy is initially added to an eikonal solver algorithm by Dellinger and van Trier [13]. Fast sweeping methods are originally proposed for isotropic media [14], however, a modification is executed to handle the anisotropic condition [15]. Single-pass or fast marching method (FMM) is another tool to compute the traveltimes, though it is not generally applicable for anisotropic medium [10].

This paper presents the initial-value ray tracing and fast marching eikonal solver in isotropic and VTI media. An anelliptic VTI wave equation is used as a kernel of the ray tracing system. The ray tracing problem solved by a forth order Runge-Kutta integrator. The fast marching finite difference algorithm, which is used as our eikonal solver, belongs to the family of upwind finite difference schemes [16]. Although the ray tracing approach is able to compute multiple arrivals with great precision, the eikonal solver is faster and more stable for traveltime computation. We discuss the result of both approaches and compare them to choose the best applicable method for the further study.

2. Theory of initial-value ray tracing

The kernel of a ray tracing algorithm is the wave equation. Fomel [17] enhanced the anelliptic qP wave approximation proposed by Muir and Dellinger [18] through replacing the linear approximation with a nonlinear one. Using the shifted hyperbola approximation, he obtained the following approximation for P wave phase velocity:

$$v_P^2(\theta) \approx \frac{1}{2}e(\theta) + \frac{1}{2}\sqrt{e^2(\theta) + 4(q-1)ac\sin^2\theta\cos^2\theta} \tag{1}$$

where $a = c_{11}$, $c = c_{33}$, $c_{ij}(X)$ are the density-normalized components of the elastic tensor, θ is the phase angle between the phase direction n and the axis of symmetry, and q and $e(\theta)$ are the anellipticity coefficient and the elliptical part of the velocity, respectively, defined by:

$$q = \frac{1+2\delta}{1+2\epsilon} = \frac{1}{1+2\eta} \tag{2}$$

and

$$e(\theta) = a\sin^2\theta + c\cos^2\theta. \tag{3}$$

Similarly, the shifted hyperbola approach is applied on the Muir's approximation to unlinearize the equation [17]. The new group velocity approximation is:

$$\frac{1}{V_P^2(\Theta)} \approx \frac{1+2Q}{2(1+Q)}E(\Theta) + \frac{1}{2(1+Q)}\sqrt{E^2(\Theta) + 4(Q^2-1)AC\sin^2\Theta\cos^2\Theta} \tag{4}$$

where $A = 1/a$, $C = 1/c$, $Q = 1/q$, Θ is the group angle, and $E(\Theta)$ is the elliptical part:

$$E(\Theta) = A\sin^2\Theta + C\cos^2\Theta \tag{5}$$

The initial-value ray tracing is broadly utilized in the shooting method of two-point ray tracing [5]. In the one-point technique, an initial given point X_0, such as the source location, and an initial ray direction P_0 are used to constitute the complete system of initial conditions for calculating the ray trajectory $X(\sigma)$ and the traveltime along the ray $T(\sigma)$. By applying the initial values $\boldsymbol{X}(0) = \boldsymbol{X_0}$, $\boldsymbol{P}(0) = \boldsymbol{P_0}$, and $T(0) = 0$ in the ray tracing system

$$\frac{dX}{d\sigma} = P \;, \qquad \frac{dP}{d\sigma} = S(X)\nabla S \;, \qquad \frac{dT}{d\sigma} = S^2(X) \tag{6}$$

rays are obtained. Using the Euler's method, an approximate solution for the ray tracing problem can be achieved by taking a small step $\Delta\sigma$ and advancing the solution according to:

$$X(\Delta\sigma) \approx X_1 = X_0 + P_0\Delta\sigma \;, \tag{7}$$

$$P(\Delta\sigma) \approx P_1 = P_0 + g(X_0)\Delta\sigma \tag{8}$$

where $g(X) = S(X)\,)\nabla S$ and $S(X)$ is the slowness function. Continuing the first step leads to the iteration

$$X_k = X_{k-1} + P_{k-1}\Delta\sigma + g(X_{k-1})\frac{(\Delta\sigma)^2}{2} \tag{9}$$

$$P_k = P_{k-1} + \frac{g(X_{k-1})+g(X_k)}{2}\Delta\sigma \tag{10}$$

and by using the third part of equation (6) and integrating a third order polynomial fitted at the end points of the $\Delta\sigma$ interval together with its derivatives, traveltime T along the ray can be defined as:

$$T_N = \frac{S^2(X_0)+S^2(X_N)}{2}\Delta\sigma + \sum_{k=1}^{N-1} S^2\,(X_k)\Delta\sigma + \frac{g(X_0).P_0 - g(X_N).P_N}{6}(\Delta\sigma)^2. \tag{11}$$

The numerical solution of the ray tracing system can be carried out, with controlled accuracy, using different techniques such as the Runge-Kutta method and Hamming's predictor-corrector method. Iterative scheme 9 and 10 are a kind of the symplectic Runge-Kutta schemes, and we use a forth order Runge-Kutta solver to determine rays.

3. Theory of fast marching eikonal solver

The fast marching method is using the fact that the direction of the energy propagation follows the group velocity equation, similar to a ray perpendicular to the wavefronts, which is defined by the phase velocity and it is called the traveltime gradient. We use equation 1 and 4 for ray tracing in locally homogeneous cells needed in this algorithm. An attentively selected order of traveltime evaluation is the main advantage of the fast marching method. While the algorithm is proceeding in a certain step, every grid point is labelled as either *Alive* (already computed), *NarrowBand* (at the wavefront, pending evaluation), or *FarAway* (not touched yet). In other words, at the beginning of the evaluation, the source positions are considered as *Alive*, and the time at these points is set to zero. A continuous band of points around the source are taken as *NarrowBand*, and their traveltime are computed analytically. Other points in the grid are marked as *FarAway* and have an "infinitely large" traveltime value [16]. The algorithm includes the following main steps:

1. Finding the point with the minimum traveltime among the *NarrowBand* points.
2. Label this point as alive.
3. Check all the immediate neighbors of the minimum point and update them if necessary
4. Repeat the procedure.

For updating, one to three neighbour points are selected which the traveltime values of them need to be smaller than the current value. After choosing the points, the below quadratic equation

$$\sum_j \left(\frac{t_i - t_j}{\Delta x_{ij}}\right)^2 = s_i^2 \tag{12}$$

should be solved for the updated value t_i. t_j are the traveltimes at the neighbouring points, s_i is the slowness at the point *i*, and Δx_{ij} is the grid size in the *ij* direction. The equation 12 is a finite difference approximation to the anisotropic eikonal equation [17]

$$v^2\left(\frac{\nabla T}{|\nabla T|}, X\right)|\nabla T|^2 = 1 \tag{13}$$

IOP Conf. Series: Earth and Environmental Science **30** (2016) 012007 doi:10.1088/1755-1315/30/1/012007

where X is a point in space, T(X) is the traveltime at that point for a given source, and v(n,X) is the phase velocity in the phase direction $n = \frac{\nabla T}{|\nabla T|}$.

4. Numerical examples

We test the above-described algorithms for isotropic and VTI ray tracing and traveltime computations in a 2D Marmousi model. The parameters that the Marmousi model (Figure 1) provides are vertical velocity V_{P0} and $\eta = \frac{\varepsilon-\delta}{1+2\delta}$. We use the slowness instead of the velocity to maximally preserve the traveltime. The study is conducted on the seismic software package of Madagascar [19]. In the first example for the initial-value ray tracing, the source is located at x=0 km and z=0 km (Figure 2), and in the second example, the source is located at x=3 km and z=0 km (Figure 3).

We trace rays in the isotropic and anisotropic condition for the both cases. Since the ray tracing theory imposes some limitations on the subsurface model, to obtain valid rays both the true and an smoothed velocity model are studied. Smoothing is carried out by applying a tridiagonal 2×2 filter for three times, and the ray window is 100° with a fan of 50 rays. It can be clearly seen that the rays traced via the anisotropic approach are well-ordered and smoother which causes a better coverage of the shadow zones than the isotropic one. Although, by smoothing the velocity, the result of the isotropic ray tracing is enhanced, the anisotropic rays of the smoothed model have a more reliable pattern and cover a wider area. We should mention that the anisotropic ray tracing is sensitive to smoothing and the smoothing parameters should not pass the threshold of the given velocity model.

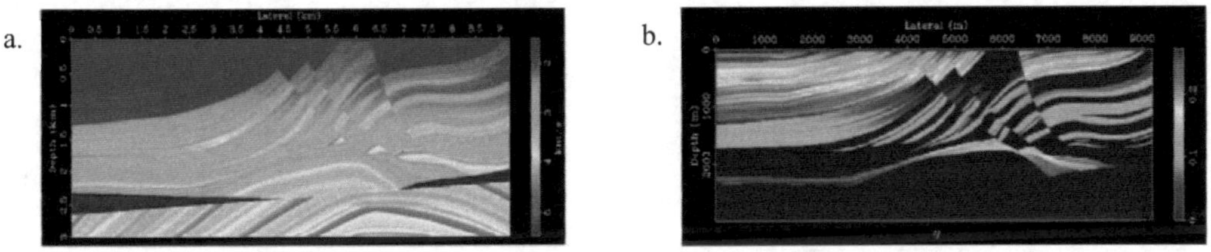

Figure 1. The Marmousi model: (a) vertical velocity model, (b) anelliptic η parameter.

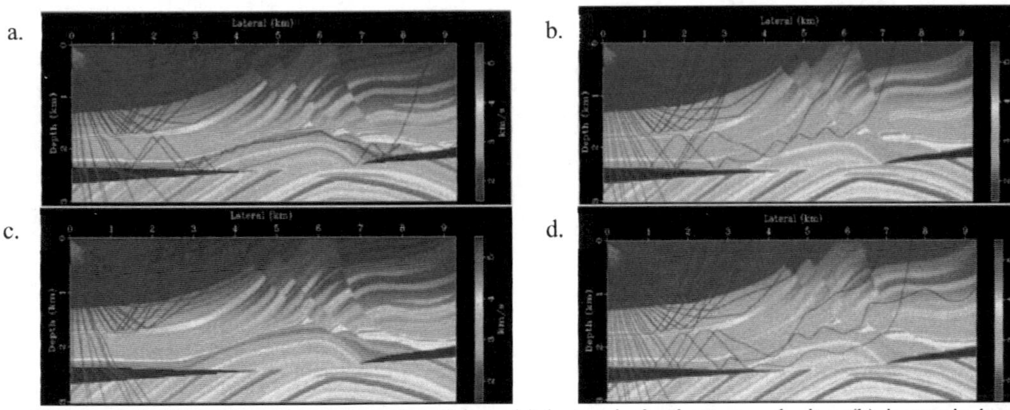

Figure 2. Initial-value ray tracing at x=0 km: (a) isotropic in the true velocity, (b) isotropic in the smoothed velocity, (c) anisotropic in the true velocity and (d) anisotropic in the smoothed velocity.

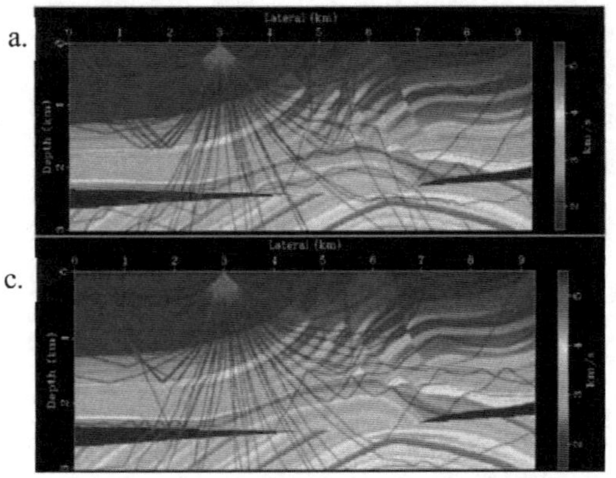

Figure 3. Initial-value ray tracing at x=3 km: (a) isotropic in the true velocity, (b) isotropic in the smoothed velocity, (c) anisotropic in the true velocity and (d) anisotropic in the smoothed velocity.

The results of fast marching eikonal solver are illustrated in Figure 4. The first arrival traveltime contours are computed in the anisotropic Marmousi model and compared with traveltime contours of the isotropic Marmousi model. It can be clearly seen that anisotropic wavefronts laterally move faster than isotropic ones, however the isotropic and anisotropic wavefronts propagate in a same speed vertically. This difference demonstrates the significance of considering anisotropy in imaging. Ignoring the shift in wavefronts position can cause error in positioning.

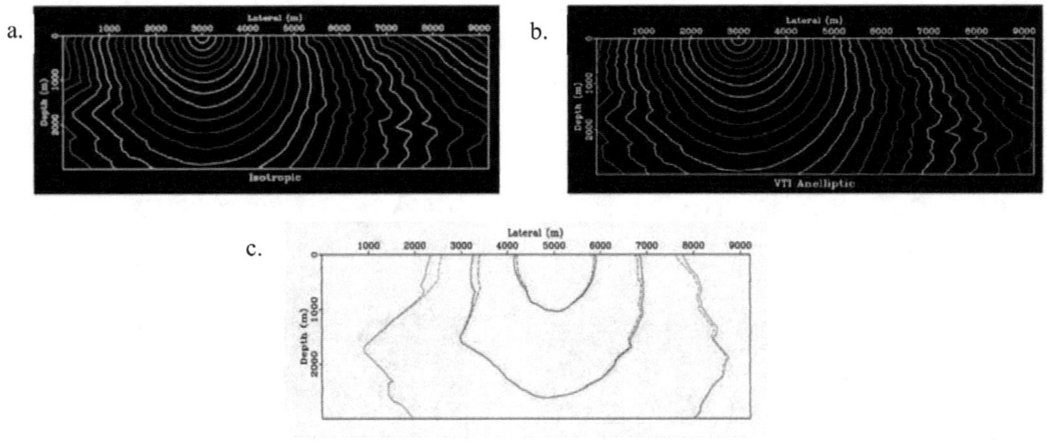

Figure 4. Fast marching traveltime contours in: (a) isotropic Marmousi model, (b) anisotropic Marmousi model. (c) Comparison of isotropic and VTI traveltimes indicates a considerable difference between isotropic and anisotropic results.

Eventually, we present a comparison between VTI ray tracing and VTI fast marching methods (Figure 5). First, the exact solution and ray tracing result are overlaid to find out the accuracy of ray tracing approach. The overlaid figure shows overlapping the ray tracing approximation and exact solution which means a great accuracy for ray tracing algorithm. Then a same procedure is performed for fast marching result which indicates an small error for contours far from the source position, however, for near distances, the accuracy is high. Finally, we overlay the contour of both approaches. We can conclude that although applying the ray tracing technique is challenging in complex velocity models and is time consuming, it remains more accurate than finite difference technique in near and far offset.

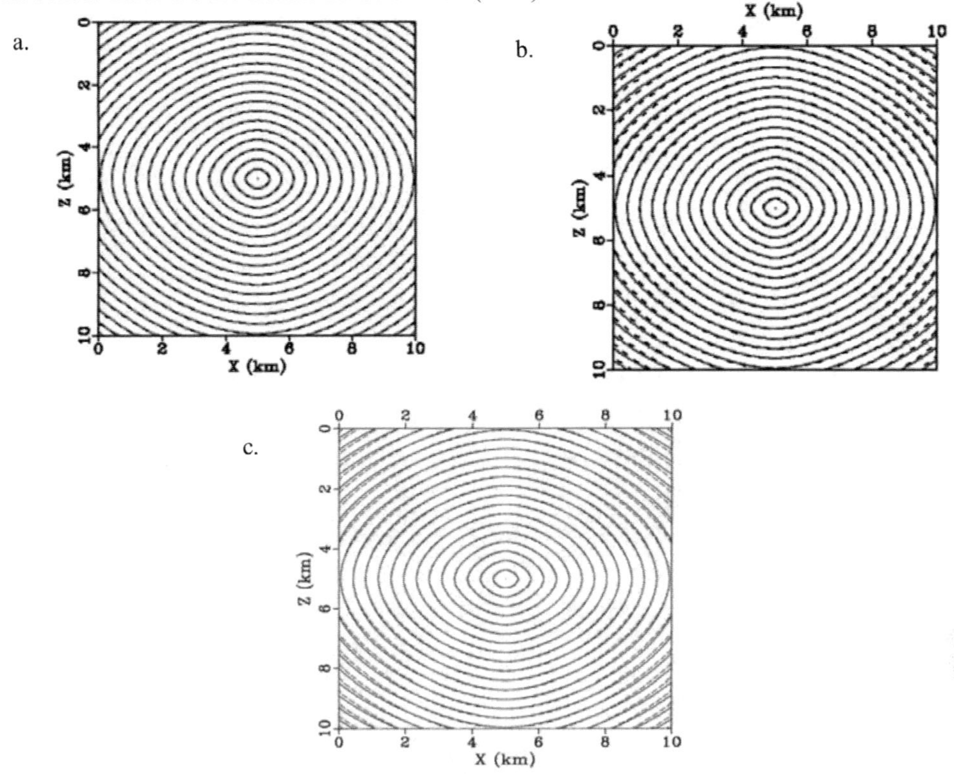

Figure 5. Comparison of VTI traveltime contours between: (a) initial-value ray tracing (solid red) and exact solution (dashed black), (b) fast marching method (solid blue) and exact solution (dashed black), and (c) initial-value ray tracing (dashed red) and fast marching method (solid blue).

5. Conclusions

We studied two techniques of initial-value ray tracing and fast marching eikonal solver in isotropic and vertical transverse isotropy (VTI) media, and we compared the results for more evaluation. The rays traced via the anisotropic approach are well-ordered and smoother which causes a better coverage of the shadow zones than the isotropic one. Fast marching method proved that anisotropic wavefronts laterally move faster than isotropic ones, however the isotropic and anisotropic wavefronts propagate in a same speed vertically. This difference demonstrates the significance of considering anisotropy in imaging. Despite the high accuracy of the ray tracing algorithm to compute the traveltime and providing the multiple arrivals, applying it in complex media is a challenging task and makes it excessively expensive for large-scale applications. On the other hand, the finite difference approach is fast and robust to define the traveltimes even in rough velocity media, though they only acquire first arrivals. Overall, since the ray tracing results are biased by the initial conditions and they are not deterministic explanations to problems, we cannot consider it as a long term solutions.

Acknowledgements

This work is funded by the grant PETRONAS Petroleum Research Fund. The authors acknowledge the contribution of PETRONAS and Universiti Teknologi PETRONAS (UTP) staffs.

References

[1] L. Behera, *et al.*, 2011 "Anisotropic P-wave velocity analysis and seismic imaging in onshore Kutch sedimentary basin of India," *Journal of Applied Geophysics,* vol. 74, pp. 215-228

[2] Z. Koren, *et al.*, 2008 "Anisotropic local tomography," *Geophysics,* vol. 73, pp. VE75-VE92

IOP Conf. Series: Earth and Environmental Science **30** (2016) 012007 doi:10.1088/1755-1315/30/1/012007

[3] U. b. Waheed and T. Alkhalifah, 2013, "Efficient traveltime solutions of the TI acoustic eikonal equation," in *75th EAGE Conference & Exhibition incorporating SPE EUROPEC 2013*.

[4] V. Babich, 1956 "A RADIAL METHOD FOR COMPUTING THE INTENSITY OF WAVE FRONTS," *Doklady Akademii Nauk SSSR,* vol. 110, pp. 355-357

[5] V. Cerveny, 2005 *Seismic ray theory*: Cambridge University Press.

[6] H. Sun and G. T. Schuster, 2003 "3D wavepath migration," *Geophysical prospecting,* vol. 51, pp. 421-430

[7] B. R. Julian, 1970 "Ray tracing in arbitrarily heterogeneous media," DTIC Document

[8] E. R. Engdahl, 1973 "Relocation of intermediate depth earthquakes in the central Aleutians by seismic ray tracing," *Nature,* vol. 245, pp. 23-25

[9] U. B. Waheed, *et al.*, 2015 "An iterative, fast-sweeping-based eikonal solver for 3D tilted anisotropic media," *Geophysics,* vol. 80, pp. C49-C58

[10] J. A. Sethian and A. M. Popovici, 1999 "3-D traveltime computation using the fast marching method," *Geophysics,* vol. 64, pp. 516-523

[11] D. Bevc, 1997 "Imaging complex structures with semirecursive Kirchhoff migration," *Geophysics,* vol. 62, pp. 577-588

[12] P. Sava and S. Fomel, 1998, "Huygens wavefront tracing: A robust alternative to ray tracing," in *SEG Technical Program Expanded Abstracts*, pp. 1961-1964.

[13] J. Dellinger, 1991, "Anisotropic finite-difference traveltimes," in *Expanded Abstracts, Society of Exploration Geophysicists*, pp. 1530-1533.

[14] H. Zhao, 2004 "A fast sweeping method for eikonal equation Mathematics of computation," *Providence, RI: American Mathematics of Society*

[15] Y.-T. Zhang, *et al.*, 2006 "High order fast sweeping methods for static Hamilton–Jacobi equations," *Journal of Scientific Computing,* vol. 29, pp. 25-56

[16] S. Fomel, 1997 "A variational formulation of the fast marching eikonal solver," *SEP-95: Stanford Exploration Project,* pp. 127-147

[17] S. Fomel, 2004 "On anelliptic approximations for qP velocities in VTI media," *Geophysical Prospecting,* vol. 52, pp. 247-259

[18] F. Muir and J. Dellinger, 1985 "A practical anisotropic system," *SEP-44,* vol. 55, p. 58.

[19] S. Fomel, *et al.*, 2013 "Madagascar: open-source software project for multidimensional data analysis and reproducible computational experiments," *Journal of Open Research Software,* vol. 1, p. e8.

IOP Conf. Series: Earth and Environmental Science **30** (2016) 012008 doi:10.1088/1755-1315/30/1/012008

Elderly Care Centre

Aliani Wagiman , Hazrina Haja Bava Mohidin, Alice Sabrina Ismail[1]
Department of Architecture, Faculty of Built Environment, Universiti Teknologi Malaysia, Skudai. 81310, Johor, Malaysia

E-mail: hazrinahaja@gmail.com

Abstract. The demand for elderly centre has increased tremendously abreast with the world demographic change as the number of senior citizens rose in the 21st century. This has become one of the most crucial problems of today's era. As the world progress into modernity, more and more people are occupied with daily work causing the senior citizens to lose the care that they actually need. This paper seeks to elucidate the best possible design of an elderly care centre with new approach in order to provide the best service for them by analysing their needs and suitable activities that could elevate their quality of life. All these findings will then be incorporated into design solutions so as to enhance the living environment for the elderly especially in Malaysian context.

1. Introduction

As mentioned by the World Health Organization (WHO), the age of an elderly is referred to 60+ years (Sim, 2002). According to the population census on elderly age group, the number rose gradually each year and it is predicted that this number will continue to grow at an accelerated rate approaching year 2020[2] (Sim, 2002).

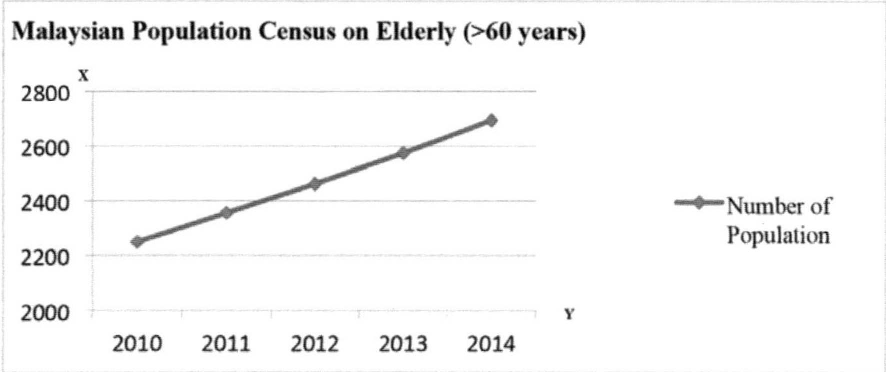

Figure 1. Malaysian population census on elderly (Source: Malaysian Centre of Statistics, 2014)

[1] To whom any correspondence should be addressed.

[2] Malaysian Centre of Statistics, 2014.

The increase is predicted on four main reasons; improved health, longer life expectancy, declining fertility and low mortality (Rashid and Tahir, 2015; Ong et al., 2009; Sim, 2002). Due to this increase, the government has come up with National Policy for the Elderly under the Ministry of Women, Family and Community Development. Most elderly have moved to the city together with their children as a result of urbanization (Malaysian Centre of Statistics, 2014). As Malaysian has become more developed, more and more elderly have decided to reside in the urban areas. Due to this massive advancement, the government has to cater for the increasing number of elderly left in the city centre be it by themselves or run by the private sector. It is a known fact that the elderly's ability in mental and physical decreases as they age (Rashid and Tahir, 2015; Ong et al., 2009; Sim, 2002). The deterioration of mental and physical ability in elderly caused some children to feel distress in taking care of their aging parents while being too busy with their everyday duties. This is the reason why more and more elderly care centre mushroomed in the city centre. This rapid growth made it essential for elderly care centre provider (public and private sector) to cater this demand (Akil et al., 2014).

In Malaysia there are two types of elderly care centre; the first is welfare elderly home for the poor which is usually state-run and the second type of care centre are the ones being catered by the NGO (Non-Government Organization). Both typologies are increasing each year as the number of elderly increases. However, the quality of these centres may not necessarily be the best provided for the elderly. This is because; today's care centres are not crafted to stimulate movements, activities and thinking which could be resolved architecturally. They are merely homes that provide a place to stay, eat and sleep. Due to this reason, these centres are usually money-centric which caused the elderly to have a mundane lifestyle adding to their already miserable lives. As a result, most of these elderly are facing mental, emotional and health problems (Tohit et al., 2012; Barnes et al., 2002).

The objective of this paper is to discuss what are the relevant architectural design approaches that could be adopted in order to identify and solve the problems of today's elderly care centre. Based on the National Policy for Elderly, architectural design approaches are best fitted into two of the most important aims of this policy. The first aim is to review the picture of ageing in Malaysia from the perspective of social security, health, community care and social services and the second aim is to review the policies and programmes for older persons, including long-term care (Akil et al., 2014; Ong et al., 2009; Torrington, 2006; Sim, 2002).

The scope of the research will be based on elderly residing in urban areas with regards to their health condition. The first group of elderly is those without complication and sickness, while the second group of elderly is those with disabilities such as intellectual impairment, immobility, senile or dementia and many more.

2. Methodology

This research will be using multiple case study method where comparison and observation will be done in two stages. The first stage will be focusing on architectural characteristics while the second stage will be focusing on the programs of selected elderly care centre. Interpretivism paradigm is chosen as the research method as it involves researchers to interpret elements. It is therefore encouraging human involvement into study of interpretation (Bevir and Rhodes, 2012). Semiotic is paired with interpretivism as it involves the study of meaning ands sign (Mohidin and Ismail, 2015). This proposed methodological framework is considered as an important contribution as it introduces new ways of looking at elderly care centre as a system of 'sign' as well as proposing various indicators in order to investigate this matter in depth. This study builds upon the theories and concepts outlined by Saussure on sign relations, Barthes on levels of signification and Gottdiener on reading the material culture as reliable ways for analysing and understanding the residential cum institutional building (Ismail, 2008). By using this framework as highlighted below, this study could be further enhanced the quality of life of the occupants staying at an elderly care centre.

Figure 2 Methodology framework

The first stage of observation focuses on spatial quality as well as form of the building while the second stage of observation focuses on adoption of suitable programs to elevate the quality of life of the elderly chosen for the study. Data collected will be used to propose the best possible design and program suitable to achieve the objective of the study.

3. Findings - Data collection – Case Study

Below are the data collected on multiple case studies:

3.1 Jan van der Ploeg, Rotterdam, Netherlands

Jan ver de Ploeg is an elderly care centre located in Netherlands. Community buildings such as a church, a school, and a playground surround it. This particular neighbourhood in an urban area which practices mix land use that incorporates commercial, recreational, community and residential areas. The centre is comprises of 79 units dormitory and composed in two complexes with 4 and 7 storeys respectively. The form of the building plays with solid and void where there is an atrium as the centre of the building. Based on figure 4, the building takes the shape of a cone. The atrium enables optimum daylighting to penetrate into the building, which brings an outdoor feeling for the occupants as well as promoting interaction between occupants, as the atrium is equipped with indoor garden facilities. There are also rooms with balconies facing inward that made it convenient for the occupants to interact freely with each other; making them feeling accompanied. Apart from providing shelter, Jan van der Ploeg offers and extensive day care program for the elderly such as nurse stations, therapeutic and podiatry facilities, not only for the elderly at the care centre but also for the surrounding community.

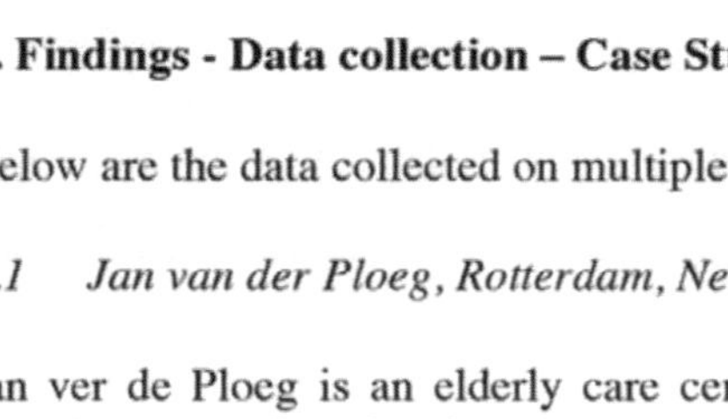

Figure 3. The indoor interaction areas at the atrium

Figure 4. Location plan of Jan ver der Ploed (Regnier, 1994)

3.2 "Säbyhemmet" Care Centre

"Säbyhemmet" is located not very far from community-centric facilities such as children day care centre, shopping mall, and residential areas. At "Säbyhemmet", there are several public interfaces that involve the community to interact with the occupants of "Säbyhemmet". The spatial layout at "Säbyhemmet" is divided into two zones. The first is the public area while the other is the private area where it is being reserved especially for the occupants. "Säbyhemmet" encourages two types of environment; first is to promote healthy surrounding between occupants and the community and the second is to be a community centre in "Säbyhemmet" itself. This is the best quality any care centre could adopt as it is stressed by Bamford and Bruce (2000), a care centre should incorporate and integrate the community into its program in order to enhance the livelihood of the elderly living in the centre. In the context of "Säbyhemmet", the children care centre is the main social contributor. These two centres work together in creating a better environment for the elderly, as the children will be the catalyst to better living for the elderly in "Säbyhemmet".

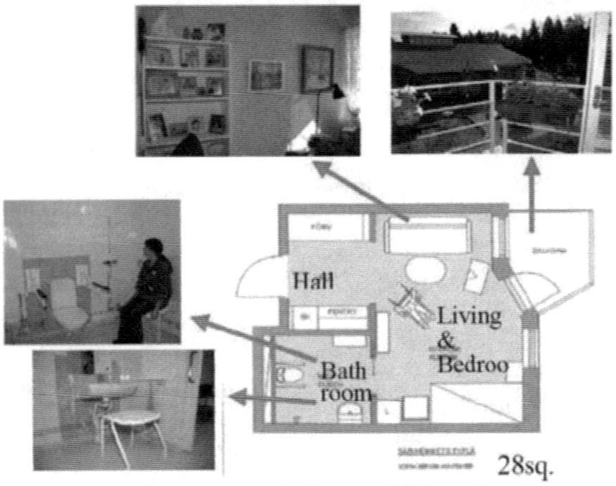

Figure 5. Spatial layout of private flats

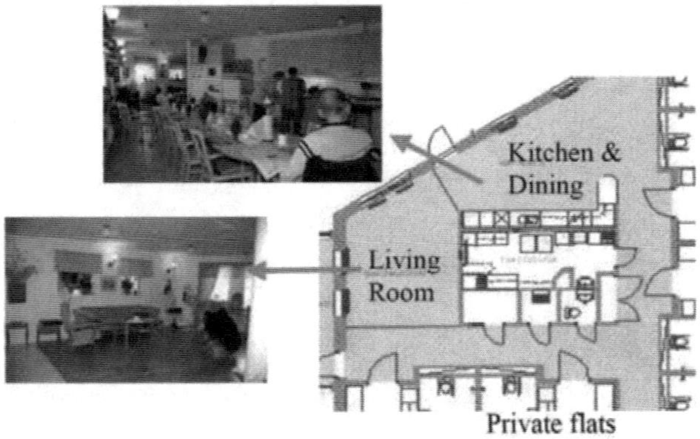

Figure 6. Public area surrounded by private flats

3.3 Al-Jenderami Elderly Care Centre

Figure 7. The frontage of Al-Jenderami

Al-Jenderami Elderly Care Centre was built in 1995 in Malaysia especially to provide care and to support the needs of senior citizens of the surrounding area. Apart from providing shelter, Al-Jenderami Elderly Care Centre also emphasized on life-long learning. The centre encourages the occupants to participate Islamic teaching classes such as reading the Al-Quran, learning how to read and write in Jawi and many more. Professional teachers are patronizing these classes. Other than that, the centre also provides full health monitoring to the occupants whenever required. The master plan layout of Al-Jenderami Elderly Care Centre is longitudinal. All the important and main spaces are being arranged in a sequence to provide a better movement and flow within the building. The centre starts off with private spaces, which includes residential areas, dining hall and followed by public interface spaces such as classrooms, mosque and a plaza.

3.4 Sri Kenangan, Cheras

Sri Kenangan elderly care centre is located in Kuala Lumpur. It is especially built to cater for people above the age of 60 years old. There are 50 occupants altogether living in Sri Kenangan. There are three category of occupants accepted in this centre, the first are those without any next of kin, second are those sent especially by the family members and the third group are homeless elderly. Sri Kenangan has special facilities such as clinic and physiotherapy as well as other therapy services apart from other necessary services such as music room, computer room, laundry room and many more.

Figure 3. Sri Kenangan entrance

Figure 9. Residential units

Figure 10. Some of the activities at Sri Kenangan

The design of Sri Kenangan elderly care centre emphasizes on sustainability architecture where natural ventilation, lighting are being applied throughout the building. It is also surrounded by greeneries and has therapeutic atmosphere.

4. Discussion

The research observed four case studies in Malaysia as well as globally. Each care centre displayed special quality of an elderly care centre. This study has made it possible to compare all four case studies and this enabled the researchers to point out the best possible design quality of an elderly care centre. However, out of the four case studies, *Jan van der Ploeg, Rotterdam, Netherlands* is the best elderly care centre as it has most of the quality needed by the occupants. Below are the findings of a good and optimum elderly care centre.

- Provide all the basic and required needs of an elderly person such as shelter, healthy food and comfortable living
- Brain enhancement programs such as classes and activities that could encourage activities
- Community and public involvement be it by adult, peer or children.
- Natural environment such as sustainable design approach and landscapes.

With all four said optimum design characteristics of elderly care centre, the needs of elderly people are taken care of. It is therefore important for future elderly care centre designers to find it crucial to adhere to all four design qualities incorporated to provide the best facilities for the elderlies.

This study is of importance for future elderly care centre designers and owners to take note in order to cater for the increasing number of elder care centres around Malaysia and globally.

5. Conclusion

It can be concluded that a good and optimum elderly care centre should provide all of the combined facilities based on the four case studies. These four traits of an elderly care centre would be the best laid out design program to be incorporated into building design suitable for a better and healthier living in an elderly environment.

References

[1] Akil, S. M. S., Abdullah, S., & Sipon, S. (2014). Challenges In Managing Elderly Care Centres In Malaysia. *International Journal of Arts & Sciences*, *7*(3), 129.

[2] Bamford, C., & Bruce, E. (2000). Defining the outcomes of community care: the perspectives of older people with dementia and their carers. *Ageing and Society*, *20*(05), 543-570.

[3] Barnes, S., & Design in Caring Environments Study Group. (2002). The design of caring [1] environments and the quality of life of older people. *Ageing and society*, *22*(06), 775-789.

[4] Bevir, M., & Rhodes, R. A. W. (2012). Interpretivism and the analysis of traditions and practices. *Critical policy studies*, *6*(2), 201-208.

[5] Ismail, A. S. (2008) *A multidisciplinary approach in understanding the symbolic meaning of built form. In: Methodology in Architectural Research*. Penerbit UTM , Johor, pp. 180-211.

[6] Mohidin, H. B. H. B., & Ismail, A. S. (2015, January). Regional Design Approach in Designing Climatic Responsive Administrative Building in the 21st Century. In *IOP Conference Series: Earth and Environmental Science* (Vol. 23, No. 1, p. 012016). IOP Publishing.

[7] Ong, F. S., Phillips, D. R., & Hamid, T. A. (2009). Ageing in Malaysia: Progress and prospects. *Ageing in East Asia: Challenges and policies for the twenty-first century*, 138-160.

[8] Rashid, A., & Tahir, I. (2015). The Prevalence and Predictors of Severe Depression Among the Elderly in Malaysia. *Journal of cross-cultural gerontology*, *30*(1), 69-85.

[9] Sim, O. F. (2002). Ageing in Malaysia, in Phillips, D. R., & Chan, A. C. (Eds.). (2002). *Ageing and long-term care: national policies in the Asia-Pacific* (No. 26). Institute of Southeast Asian Studies.

[10] Tohit, N., Browning, C. J., & Radermacher, H. (2012). 'We want a peaceful life here and hereafter': healthy ageing perspectives of older Malays in Malaysia. *Ageing and Society*, *32*(03), 405-424.

[11] Torrington, J. (2006). What has architecture got to do with dementia care?: Explorations of the relationship between quality of life and building design in two EQUAL projects. *Quality in Ageing and Older Adults*, *7*(1), 34-48.

IOP Conf. Series: Earth and Environmental Science **30** (2016) 012009 doi:10.1088/1755-1315/30/1/012009

Tree Water Uptake And Suction Distribution On Tropical Residual Soil Slope

M. F. Ishak[1], F.F. Sulaiman[2], N. Ali[2], A. Kassim[2]
[1]Faculty of Engineering Technology, University Malaysia Pahang.
[2]Faculty Of Civil Engineering, Universiti Teknologi Malaysia.

E-mail: fakhrurrazi@ump.edu.my

Abstract. This paper present an exploration of soil matric suction effected by water uptake via tree root at toe of slope on various condition between wet condition (high rainfall) and dry condition (prolonged no rainfall). Matric suction generated by active root tree has substantial influence soil moisture content on residual soil slope. A field monitoring was carried out to collect matric suction data at slope in two conditions; with a tree located at toe of slope and absent of a tree. The installations of instruments particularly at slope with tree at toe were placed within vicinity of the tree with certain depths and distances. The matric suction data from field monitoring was influence by the rainfall events that lead to the instability of soils slope. Analysis of soil matric suction distribution pattern indicates that the highest matric suction value was at shallower depth and proximity of tree. The matric suction profiles obtained from field monitoring are applied as an input data to develop soil matric suction contour. The effect of transpiration driven by active root zone generated matric suction on soil at vicinity of tree may create dry soil to increase soil shear strength.

1. Introduction

The studies on the side slopes of a railway embankment that began to fail after several years of construction due to the gradual loss of stability resulted from a decrease in the strength of the soil with time has been done by Krahn *et al.* (1989). Slope failures commonly occur when the shear strength of the soil is reduced through a decrease in effective stress due to pore water pressure increment (Glendinning *et al.*, 2009). The field measurements showed that there was a decrease in suction near slope surface. There is evidence that the suction may have been at or near zero at the point of instability.

This study illustrates the important role of matric suction that play in the stability of slopes, particularly in the near surface stability. Related to that, matric suction produced by vegetation plays an important parts for slope stability by providing immediate shear strength enhancement and modifying the saturated soil water regime, which is turn causes a variation in soil suction or pore pressure (Simon & Collison, 2002). Vegetation also can enhance the stability of a slope by root reinforcement.

Many researchers (Thorne, 1990; Ali, 2010; Normaniza *et al.*, 2007) have discovered the effect of mechanical on tree root that can benefit in preventing shallow slope failure but only a few studies to quantify the hydrological effect for the potential benefit stabilization of slope. Apart from providing natural mechanical soil reinforcement, tree roots dissipate excess pore water pressure and produce sufficient matric suction to increase the shear strength of the surrounding soil. The hydrological effect is related much closed to soil moisture variation and can be directed through transpiration. This effect is found to be important as mechanical effect provides significant increase in soil strength that will definitely improve slope stability in certain conditions. It must be considerate before cutting and felling down the trees without understanding the hydrological condition that have a great impact on the stabilization of slope and soil surrounding it.

Lim *et al.* (1996) and Hossain (2010) has stated that, variations in matric suction profile have been caused by changes in climatic conditions with time. However, the presence of vegetation has shown a significant increase in the in situ matric suction and changes the total head profile and soil in that area (Lim *et al.*, 1996). The hydrological effect is related much closed to soil moisture variation and can be directed through transpiration.

The field measurements of matric suction changes were observed at slope with and without tree at toe to provide useful data for analysis of study area at Faculty Electrical Engineering, Universiti Teknologi Malaysia. The objective of this study is to analysis the effect of tree induced suction at various conditions from wet condition (high rainfall) as an initial condition to dry condition (prolonged no rain). The soil suction results were adopted to develop soil suction contour influence by tree at toe of the slope. This study focuses on the pattern of soil suction that generate within the vicinity of a single representative mature tropical tree.

2. Material And Method

This study was carried out during field monitoring period from 4th February 2012 to 15th February 2012. The tree located at the toe of slope in front of P16 at Faculty of Electrical Engineering, Universiti Teknologi Malaysia as showed in Figure 1. The *acacia mangium* tree was preserved at toe of the slope and no irrigated during this study period to avoid complication of water wetting to soil proximity of the tree.

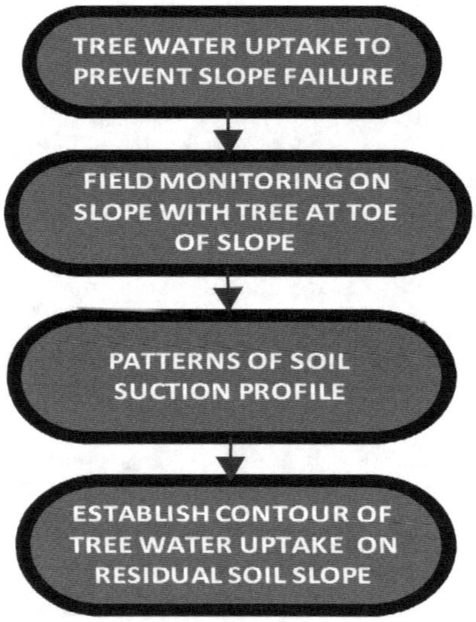

Figure 1. Flow chart of research stages

Field monitoring was carried out to determine the in-situ matric suction on the slope with and without tree at toe. Throughout this field monitoring, the matric suction measurements were taken manually. The installed instruments are concentrated at single mature tree at toe of slope. Two types of soil suction measurement instruments were used in the study, i.e. tensiometer and gypsum block. These combinations of instrumentations were placed at Slope Station and Flat Station as shows in Figure 1 and 2, respectively. Each station of tensiometers were placed at distance of 0.1*h*, 0.2*h*, 0.4*h* etc (where *h* is the height of the tree), according to Biddle (1998). Then, the instruments were installed according to arrangement of 1m, 2m and 4m distance from tree trunk with depth of 0.5m, 1.0m, 1.5m and 2.0m.

To monitor soil suction changes of low suction value at slope with *acacia manggium* tree at toe, Jet-fill tensiometers were used. Soil suction range between 0 kPa to 100 kPa is measured directly by Jet-

IOP Conf. Series: Earth and Environmental Science **30** (2016) 012009 doi:10.1088/1755-1315/30/1/012009

fill tensiometer. A tensiometer comprises of a tube with a porous ceramic tip on the bottom, a vacuum gauge near the top and a sealing cap.

To measure the high suction value beyond 100kPa up to 1000kPa, the Gypsum block with a measuring capacity range of 10kPa to 1500kPa was introduced. The Gypsum block is cylindrical having a diameter of 22 mm and a length of 24 mm. This instrument consists of two parts; the gypsum soil blocks (sensor) and the measuring instrument Moisture Tester. At this study area, the gypsum blocks and tensiometer were installed with same arrangement throughout both Station (Slope Station and Flat Station).

2.1. Residual Soil Profile
In this study, the subsurface investigation was conducted to determine and identify the thickness and various distribution materials of weathering zones in residual soils at Faculty of Electrical Engineering. Data collected from the trial pit with maximum depth of 1.5 m was analysed to produce the generalized soil profile characteristics of the residual soil at the study area.

Through this site investigation, it is indicated that the soil was formed in-situ by the decomposition of parent material, which the trace of uplifted and transportation marks cannot be found in this study area.

3. Results And Discussion

3.1. Soil Properties
The disturbed and undisturbed soil samples were collected at the ground surface up to 1.5 m depth of the study area. A series of laboratory testing were conducted to determined soil properties, which can determine the soil type and geotechnical properties as presented in Table 1. The main physical index property of the soils investigated in this study was soil classification, which depends on several factors such as the Atterberg limits, specific gravity, particle size distribution, porosity and void ratio. This soil formed in-situ under tropical weathering condition as defined by The Public Work Institute Of Malaysia (1996). Based on BSCS classification, the soil is classified as in grouped of sandy SILT (MVS)

Table 1. The properties of the soil material in this study area

Composition	Sandy SILT
Gravel (%)	5.1
Sand (%)	20.9
Silt (%)	48.7
Clay (%)	25.3
Specific Gravity (G_s)	2.62
Void Ratio (e)	1.44
Porosity (n)	0.59
Permeability (k_{sat} (m/s))	4.1×10^{-7}
Effective Cohesion (c')	9
Effective Friction Angle (ϕ')	23
Unsaturated Friction Angle (ϕ^b)	20

3.2. Soil Water Characteristic Curve (SWCC)

The soil-water characteristic curves (SWCC) of residual soils at Faculty of Electrical Engineering Universiti Teknologi Malaysia are shown in Figure 3. The SWCC for low suctions (less than 1500kPa) was determined by fitting the average value from a series of pressure plate extractor tests. Based on this SWCC, the parameters such as saturated volumetric water content (θ_S), air entry value (A_{ev}) and residual volumetric water content (θ_r) of the soils can be identified and shown in figure 2.

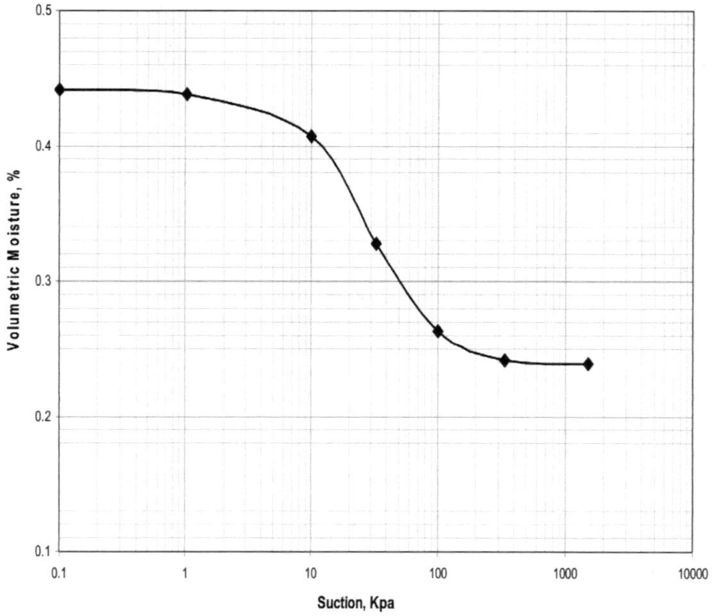

Figure 2. Soil Water Characteristic Curve (SWCC)

3.3. Soil Suction Profile

Soil suction profiles related to soil depth at Slope Station and Flat Station on 4th February 2012 and 15th February 2012, are shows in Figure 3,4, 5 and 6, respectively. During this field monitoring, dry period was occurred of 9day without rainfall was selected to represent condition of suction profile due to tree water uptake. First of all, Figure 3 and 4 shows the matric suction profiles right after prolonged and antecedent rainfall on both slopes (slope with and without tree at toe). Also, soil suction profile on 4th February 2012 was identify as lowest value regarding to antecedent rainfall occurred for 4days and indicate as an initial condition for the analysis.

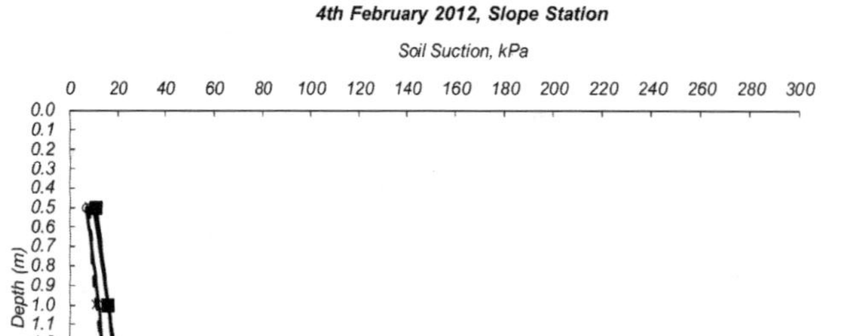

Figure 3 Soil suction profiles on 4th February 2012 at Slope Station

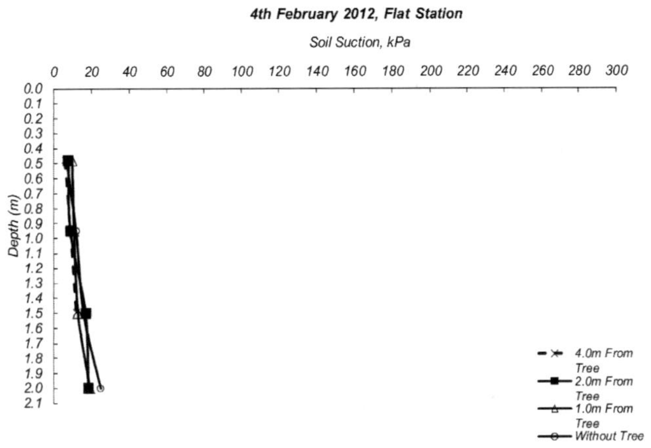

Figure 4 Soil suction profiles on 4th February 2012 at Flat Station

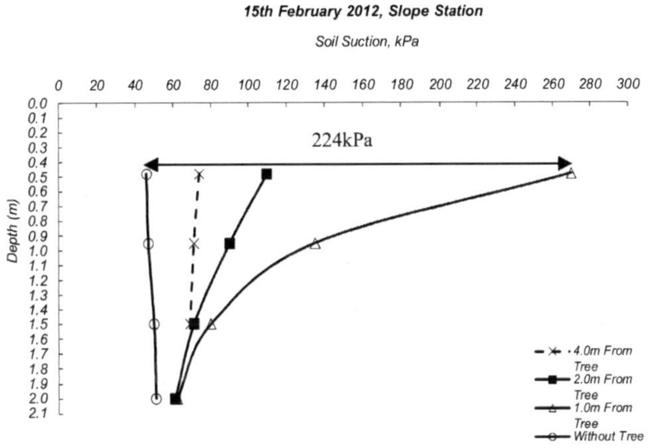

Figure 5 Soil suction profiles on 15th February 2012 at Slope Station

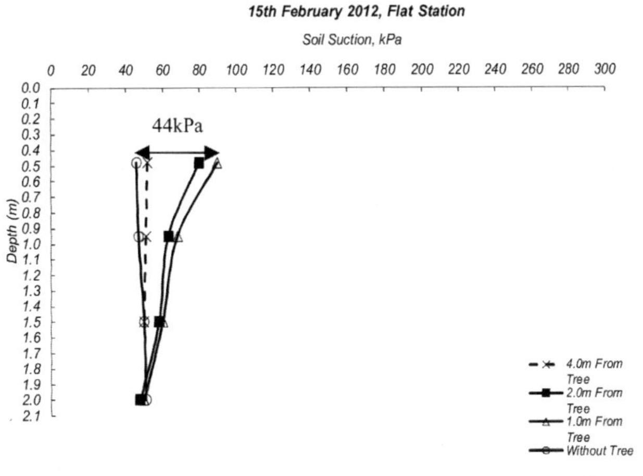

Figure 6 Soil suction profiles on 15th February 2012 at Flat Station

Soil suction value at Slope Station with and without tree at toe on 4th February 2012 and 15th February 2012 was shows in Figure 4 and 6 respectively, has revealed a significant increase in matric suction. This similar pattern of suction was showed at Flat Station in Figure 5 and 6. While, the highest suction value was encounter on 15th February 2012, the maximum value was occurred at both Stations due to absent of rainfall for 9days. Furthermore, suction was gradually decreased at distance of 2.0m and continues to decrease at distance of 4m from tree. The soil suction distribution patterns of changes were developed with different condition or weather can be clearly seen. The matric suction at near surface particularly at 0.5m depth increase gradually to the right as showed in Figure 6 and 7. These figures also indicate that the matric suction at 1m from tree (at Station Slope and Station Flat), particularly at depth of 0.5m, which higher than at slope without tree up to 244 kPa and 44 kPa, respectively.

It can be concluded, the soil suction value was significantly increase within vicinity of tree and decrease subject to distance from tree. It also discover that suction value was significantly difference occur at particularly swallow depth of tree. This is owed by active *acacia mangium* tree root zone that produce sufficient suction on soil compared to the soil suction without tree (Biddle, 1998).

3.4. Soil Suction Contour

The result variation of the soil suction encounter during field monitoring was show in Figure 5 and 6 are applied as an input data to develop soil suction contour. Figure 7 show the soil suction contour on 15th February 2012, the highest suction value of 240kPa was generated at the base of tree near the toe of slope and the lowest suction value of 42kPa as suction without influence by tree. The effect of tree induced suction on soil was concentrated at preserved tree, while the suction depleted when distance away from tree.

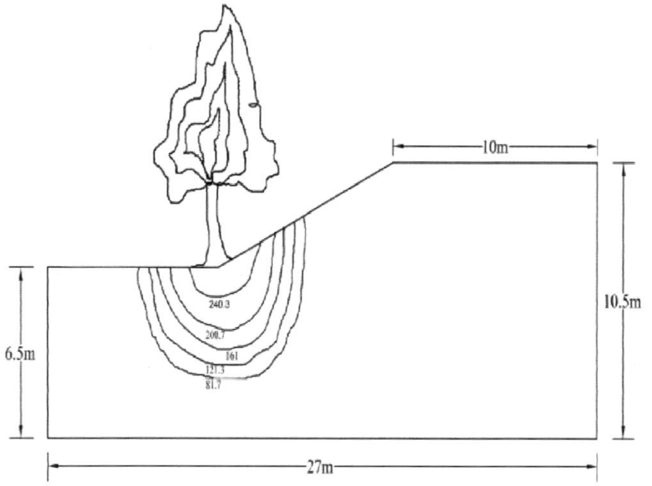

Figure 7 Matric suction (kPa) contour on 15th February 2012

The comparison of matric suction between with and without trees was done by previous researcher such as Pollen-Bankhead *et al.*, (2010) and Woon *et al.*, (2011). In their comparison of matric suction developed in vegetated and bare soil columns in the field and physical model in the laboratory, it was found that almost 100% of matric suction generated in the vegetated column higher than bare soil columns. In relation to that, the measured matric suction at slope without tree is attributed by evaporation which is lower than at slope with tree at the toe, as it was induced by transpiration. The soil suction contour as show in Figure 8 exposed the soil suction at the toe is higher than other section. This contributed by *acacia mangium* tree and helps enhance the soil matric suction to pull out water from soil via active root tree in that area.

4. Conclusion

Weather condition is one of factors that influence soil suction variation at this study area. From the overall profiles results showed that tree water uptake influence soil suction distribution on slope with tree at toe and can be summaries as follow:

i. The results showed that water uptake by tree root mostly influenced by the weather of antecedent rainfall and dry period. The soil suction near the ground surface is more sensitive to the variation of weather, followed by those at greater depths.

ii. The results disclose that tree water uptake caused soil suction increased significantly at vicinity of tree and reach high soil suction value than soil suction without tree. Maximum changes in suction occur at close proximity to tree trunk and near to ground surface but the changes in soil suction decreases gradually with depth and distance from tree. It shows that, increased of soil suction was driven by tree water uptake at active root tree zone.

iii. This study revealed the matric suction generated by root tree has significantly increased values that presented in soil matric suction contour for vertical and radial direction from the tree between 9days of this analysis. Contribution of single mature tree has significantly altered the soil suction or soil moisture distribution trigger by transpiration on unsaturated soil slope.

References

[1] Ali, F. 2010. Use Of Vegetation For Slope Protection: Root Mechanical Properties Of Some Tropical Plants. *International Journal of Physical Sciences*. Vol. 5(5), 496-506.
[2] Biddle, P.G. 1998. *Tree Root Damage To Buildings*. Wantage, Willowmead Publishing Ltd.
[3] BS, 1990. *Methods of Test for Soils for Civil Engineering Purposes, (BS 1377: Part 1-9)*. London: British Standards Institution.
[4] Glendinning, S., Loveridge, F., Starr-Keddle, R.E., Bransby, M.F. and Hughes, P.N. 2009. Role Of Vegetation In Sustainability Of Infrastructure Slopes. Geotech. Eng., Proc. Inst. Civil Eng. 162, 101-110.
[5] Hossain. M. K. 2010. Effect of rainfall on matric suction and stability of a residual granite soil slope, DUET Journal, Vol. 1, Issue 1.
[6] Krahn. J., Fredlund. D.G., and Klassen M.J. 1989. Effect of soil suction on slope stabilityat notch hill, Canadian Geotechnical, Vol 26:269-278.
[7] Lim. T. T., Rahardjo. H., Chang. M. F., And Fredlund. D. G. 1996. Effect of rainfall on matric suctions in a residual soil slope, Canadian Geotech Vol. 33(4): 618 – 628.
[8] Normaniza O., Faisal, H.A. and Barakbah, S.S. 2007. Engineering properties of leucaena leucocephala for preventing of slope failure. *Journal Ecological Engineering*, 32 (2008): 215-221.
[9] Pollen-Bankhead, N., and Simon, A. 2010. Hydrologic and hydraulic effects of riparian root networks on streambank stability: is mechanical root-reinforcement the whole story?, Geomorpology, 116(3-4), 353-362.
[10] Public Works Institute Malaysia 1996. *Tropical Weathered In-Situ Materials*. Geoguides: 1-5.
[11] Simon, A. and Collison, A.J. 2002. Quantifying the mechanical and hydrologic effects of Riparian vegetation on stream- bank stability, Earth Surface Processes and Landforms, 27: 527–546.
[12] Thorne, C.R. 1990. *Effects of vegetation on riverbank erosion and stability. In Vegetation and Erosion: Processes and Environments*, John Wiley and Sons, 125–144.
[13] Woon K.X., Leung A.K., Ng C.W.W., Chu L.M. and Hau B.C.H. 2011. An Experimental Investigation On Suction Influence Zone Induce By Plant Transpiration. *Unsaturated Soils: Theory and Practice. Proceeding of the 5 Asia-Pacific Conference on Unsaturated Soils, Dusit Thani Pattaya, Thailand*. Vol II, 861-866.

A Review of Economic Factors Influencing Voluntary Carbon Disclosure in the Property Sector of Developing Economies

J U Kalu, G U Aliagha and A Buang.

Department of Real Estate, Faculty of Geoinformation and Real Estate, Universiti Teknologi Malaysia, 81310 UTM, Skudai, Johor Malaysia,

Email: khalujosef@gmail.com

Abstract: Global warming has consequences on the environment and economy; this led to the establishment of United Nation Framework Convention on Climate Change (UNFCCC) and the Kyoto Protocol. These two agreements were to reduce greenhouse gases (GHG) emissions which are responsible for climate change and global warming. Developing countries under the protocol are not obligated to reduce or disclosure GHG emission, so their participation in the protocol is on voluntary mitigation bases. This study intends to examine economic factors that influence voluntary carbon disclosure in the property sub-sector of developing countries based on annual report of listed property companies in Malaysia. Signaling theory addresses the problem of information asymmetry in the society. Disclosure is an effective tool to overcome information imbalance among different market participants. The study hypothesizes that the economic factors that influence voluntary carbon information disclosure in developing countries are: [1] the company's size; this is because a large-sized company have more resources to cover the cost of reducing pollution. [2] The company's gearing status; where there is no sufficient information disclosure in a highly geared company will result to an increased agency cost. [3] Profitability; profits grants companies a pool of resources for mitigation activities and environmental reporting. Also, carbon disclosure acts as a means for achieving public confidence and legitimacy. [4] Liquidity: Companies that are highly liquid will disclosure more information to distinguish themselves from other companies that are less liquidity. This is correlated to environmental disclosure. [5] Financial slack affects companies' ability to participate in green technology projects that enable a reduction in emission.

Keywords: Carbon emission, greenhouse gases, Property sector.

1.0 INTRODUCTION

Changing landscapes, sea level rise, severe storms and increased risk of flooding and drought - these are some the examples of the effects of climate change [1]. Climate change has serious consequences on all levels of life on earth; especially in the environment and society [2-5]. However, there is also the impact on the national and global economy, with the negative consequences, ranging from loss of earnings [6, 7], to climate change-related disasters[8]; with a subsequent increase in cost of emission reduction. The result of climate change led to the establishment of two agreements to address the pressing issue of reducing anthropogenic greenhouse gases (GHG) emissions in the atmosphere: the United Nation Framework Convention on Climate Change (UNFCCC) and the Kyoto Protocol. The UNFCCC is a joint treaty designed at alleviating GHG concentrations in the atmosphere at a level that would prevent dangerous anthropogenic interference with the climate system. The Kyoto Protocol came into being on 11 December 1997 at the third Conference of the Parties (COP) of the UNFCCC, and entered into force on 16 February 2005. The purpose of the Protocol is to encourage countries to reduce the emissions of carbon dioxide and other gases which cause greenhouse effect. The Kyoto protocol requires developed countries to reduce their emissions of GHG by a collective average of 5% below their 1990 levels [9]. These regulations became necessary due to the activities of companies that are main sources of emissions responsible for the greenhouse effect. This is because industries all over the world depend heavily on the burning of fossil fuel for manufacturing activities and power generation, hence the main contributors to global warming and climate change. While in the past companies, despite their contribution to greenhouse gas emissions have been largely excluded in the search for solutions, now carbon disclosure is a mechanism for market stakeholders to make an informed decision on the state of a company's emissions performance, and exert pressure on companies to reduce their emissions. Developed countries that ratified the Kyoto protocol are obligated to enact regulation incorporating the protocol's provision on disclosures while developing countries are not statutorily required to reduce their emissions under the protocol. Therefore mitigation and disclosure in developing economies are voluntary. So firms in developing economies are not bound to mandatory carbon disclosure, like firms in developed countries that have ratified the protocol. While industrialized counties set targets for significant reduction in CO_2 equivalent emissions by 2010, it was discovered that the cheapest source of CO_2 emission reduction is not with the industrialized nations but in developing countries. So it is accepted that for Kyoto protocol to achieve its objective of slowing global warming it will need the participation of developing countries in accepting an emission reduction commitment. This study tends to examine the economic factors that influence voluntary carbon disclosure in the property sub-sector of developing economies. The study is focusing on annual reports of listed property companies in Malaysia.

2.0 CARBON DISCLOSURE FOR CLIMATE CHANGE MITIGATION

The overall, the average temperature has increased by about 85°C since 1880 [10]. This has led to the experience of an upsurge in the surface temperature and the concentration of CO_2 in the atmosphere and the sea level is expected to increase from 26 to 81 centimeters by the year 2100 [10]. It is most likely that humans are the dominant cause of global warming, and without emissions reduction, the world is on track to a more than 2°C or 4°C possible warming by the end of this century [10]. Thus, there is a growing demand for the disclosure of information related to carbon [11] ;[12] especially within companies in the developing countries.

The burning of non-renewable energy sources, deforestation, farming, and transport result in the discharge of greenhouse gases (GHG) into the atmosphere. Greenhouse gases such as carbon dioxide (CO_2), methane (CH_4), nitrous oxide (N_O) and chlorofluorocarbons (CFCs) are considered to be the main constituents to the greenhouse gases, which is the main cause for global warming. Research has shown that in the past 150 years human commercial activities have brought up CO_2 levels from 280 ppm to 379 ppm. Although some dissenting observations exist, researchers has gradually arrived at an agreement that the accumulation of these gases can change the environment of the Earth, with serious effects on human health and the environment, as well as significant threats to businesses. Many developed and developing countries have identified the risks created by global warming and are taken steps to mitigate the greenhouse effect, as part of their reaction.

Global warming is an issue that impacts a variety of companies around the world. The effects go beyond the ecological scopes, being connected to energy security and performance. It has become a subject of social, regulating and economic attention recently, and has been brought to the front as an "inconvenient truth" that requires a serious policy strategy [13]. While an obligatory control is unavoidable, organizations have been willingly disclosing information about their emission [14]. Although industry-focused strategy to climate change has taken over the public conversation and significantly supports the development of sustainability report.

2.1 VOLUNTARY CARBON DISCLOSURE IN THE PROPERTY SECTOR

Given the complexity of climate science and the impact it has on the overall well-being. The task of the state policy on climate change is huge and policies around carbon minimization is emerging around the world [14]. In 2009, the Climate Change Conference of the United Nations in Copenhagen (COP15) discussed future commitments to minimize carbon and climate change mitigation.

Carbon Disclosure has become a significant establishment for governance, awareness on climate change, clean energy and energy efficiency, and legitimacy. Companies measure the carbon emissions at the request of interested parties. Carbon measurement is done to get ahead of competition and regulation; or simply to identify spending cuts, there are a number of reasons to report this information voluntarily. Emissions information disclosure and reduction targets help to build trust among stakeholders and investors, and can help businesses understand their performance in the industry.

Carbon Disclosure has grown exponentially in recent years. This may be connected with the awareness of climate change and its impact on business and society as a whole, in tandem with constant demands for accountability from the investment community. Countries around the world are developing policies, particularly for the protection of the environment.

Carbon Disclosure can be understood as developing institution which provides a level of management in the broadest sense, as a multi-level, multi-actor system in regulations and standards[15, 16]. The role of markets [17] are also emphasized and the operational solutions and strategies that are important aspects of the company's management. Carbon Disclosure plays a role in three control modes; compliance, carbon trading and management through standards and civil accountability and transparency. Disclosure of carbon is important for regulatory compliance. It is also part of the carbon markets and carbon Management. Carbon markets are a key form of government, by imposing restrictions on emissions and put a price on carbon. Indeed, carbon as a tradable commodity is irrelevant built only on specific information protocols that require the development of systems of accounting for carbon [18]. Companies can also participate in the voluntary carbon markets, by investing in clean development mechanism (CDM) in developing countries.

Many companies are implementing carbon management systems, sometimes in the form of modules related to their accounting systems and logistics control. A key factor is the prospect of increasing internal efficiency, to reduce energy costs and calculation of the carbon footprint at the product level. Increasingly, companies push their suppliers to participate in the carbon accounting for the purposes of labeling and controlling costs.

Disclosure of carbon information helps to put pressure on the organization for accountability and transparency, the pressure which is [19] termed "civil right" is enforced by non-government organizations. Most companies are beginning to provide a very informative analysis of the financial risks of climate change. The financial community is very focused on determining which companies are threatened by climate legislation, this type of exposure, and how it affects the potential investment.

2.2 Evolution of carbon disclosure

The evolution of carbon disclosure demonstrates the importance of creating a "business model" that is institutionally viable. As a result, business leaders have expressed doubts about the value of investments in external communications in the absence of the public [20]. The establishment of measurement and reporting of carbon are still growing. Businesses are becoming increasingly interested in the carbon information systems that are useful for measuring and monitoring greenhouse gas emissions at the facility, process or product level. Carbon Disclosure, therefore, has potential value to managers and companies. Over the years most firms and governments have created carbon and clean energy practices, for these companies, the carbon measurement, management and reporting, and analysis represent a vast new opportunity in the market.

Measurement for carbon is now considered an important stream of research that attracts a growing community of international researchers [9, 21]. Previous studies have generally focused on the economic impact of carbon emissions [22] or incentives for the management of voluntary disclosure of GHG [23, 24]. In particular, some studies have examined the influencing factors at the country level and found that ratification of the Kyoto Protocol [25], the severity of the regulatory environment, the environmental sensitivity of the private, market structure [26] and the resolutions of shareholders [27] are crucial in disclosing the company's GHG emissions. Most of these carbon studies have focused on developed countries; this study looks at the developing countries.

Declaration of carbon is part of global efforts to reduce carbon emissions, which include substantial investments and long-term commitment. Reporting often requires the creation of a carbon emissions inventory and implementation of carbon management. Also, once a company launches to the public disclosure of carbon, it is expected that the company will continue to do so, and termination of the disclosure can be considered by the market as a signal of lack of knowledge on climate change or the company's inability to manage the risk of carbon, which eventually may set a negative reputation with stakeholders [28]. Thus, the lack of financial resources does not allow us to successfully adapt to internal and external pressure to change [29].

Companies that disclose their greenhouse gas emissions have a number of reasons, such as to maintain the legitimacy to defend its reputation to get the right picture and to participate in sustainable development programs voluntarily [12]. A couple of studies [23, 25] attempted to analyses the factors that affect voluntary disclosure of greenhouse gas. Because disclosure is not mandatory, it is likely that the limited resources outweigh the needs of the stakeholders. Thus, we expect that companies with higher leverage disclosure more. In companies with high growth potential, economic goals will be a high priority than environmental reasons [23, 30]. Thus, these companies will prioritize their actions to achieve financial goals, resulting in fewer resources for actions to reduce carbon emissions and disclosure. We expect that limited resources will be more pronounced for companies in developing countries.

2.3 THE EFFECTS OF CLIMATE CHANGE IN MALAYSIA

Malaysia is ranked 87th among 180 countries in terms of the degree of climate risk [31]. The real estate sector was hit particularly hard. Historical records show that Malaysia has experienced severe flooding in 1926, 1963, 1965, 1967, 1969, 1971, 1973, 1979, 1983, 1988, 1993,1998, 2005, and December 2006 to January 2007 double floods (occurred in the state of Johor). It is estimated that the average annual flood damage in Malaysia to be about RM 3billion (USD 0.79 billion) [32, 33]. The flooding of January 1971 had resulted in a loss of more than RM 200 million (USD 32.8 million) and the death of 61 people. In 2006-2007 the Johor state massive double flood had incurred a total cost of about RM 1.5billoion (US $ 0.41billion) and the death of 18 people and about 110 000 people evacuated and sheltered in relief centers. Furthermore, Peninsular Malaysia has more than 29,000 km2 of the total area that is prone to flooding [34, 35]. The trend of increased frequency and intensity of rainfall and rising sea level has the potential to expand flood zones and increase the problems of flood management, waves, and the erosion of shoreline.

It is anticipated that as 2020 approach, Malaysia should anticipate serious challenges in flood and drought management that will come with increased severity and frequency of flash floods, prolonged droughts, also with the threat of the corridor area's exposure to 189 river basins [36]. As flooding results mainly from heavy rainfall during the monsoon seasons, the effect of climate change begins to manifest, as evidenced by the unprecedented flood of 2006-2007 state of Johor [37, 38]. This study looks at the participation of the property sub-sector in climate mitigation and environmental disclosure in general and carbon disclosure in particular in the annual reports of listed property companies in Malaysia.

3.0 THEORETICAL FRAMEWORK

3.1 Signaling Theory

Voluntary reports are efforts to eliminate the asymmetry of information between the company and external suppliers and investors in the investment community. "The concept of the voluntary disclosure provides that companies that have excellent environmental performance do not cover the environmental effects of their duties but voluntarily inform stakeholders about their actions in the field of sustainable development. The voluntary disclosure provides risk information to current and potential investors [39]. Voluntary disclosure can lead to a competitive advantage. When disclosure is absent or low stakeholders will assume the current environmental strategy adopted by the firm is inferior [40, 41]. Superior environmental performers truly disclose issues regarding environmental affairs, the quality of their disclosures are superior to the quality of the weak environmental performers. The superior firms believe that their strengths will outweigh the weaknesses and do not fear the reaction of any stakeholder [40].

Information signaling theory addresses the problem of information asymmetry in the society. In social research, the idea is based on the training work [42], which used the idea of the labor market, to show how candidates indicated potential to organizations. It was stated that a [42] high-quality job applicants signal through high education papers and resumes distinguishing themselves from low-quality applicants. Thus, the employers lack of information about the quality of the applicants as such the applicants acquire an education to signal their quality and reduce information asymmetry which often hampers the selection ability of the prospective employers. In this instance, education is used as a means of communicating the unobservable characteristic of the job applicant [43].

The designation will be an iterative process, which has also obtained most expensive cost of signaling [44]. There are different ways in which organizations can specify information about itself. Of this voluntary disclosure of useful information, reporting is regarded as one of the most profitable [45]. Signal theory can also be used to achieve the objectives of other concepts, [46] and it is recommended that, the theory can be widely used in the social sciences. It is also the opinion of a study [46] that the theory when focusing on desired signal indicate that the part of the signaling random negative information is neglected. It is [47] recommended that, in practice, parties should send various alerts, not knowing the report that could have a beneficial effect. In addition, the predictive capacity of the theory depends on the assumption that the recipient of the information, understands perfectly what is disclosed by the sender, yet the nature of the operating environment shows that evidence of a large company may affect the interpretation of the recipient [46].

Disclosure can be an effective way to overcome the asymmetry of information between the different market participants. In the context of climate change, many decisions must be made by consumers, investors, entrepreneurs and politicians, who will determine the success of the transition to a low carbon economy. Reports on energy and carbon information have increased over the past two decades, climate change and the role of business in their resolution became clearer. Measurement standards have emerged, but for most, it is voluntary for companies to disclose their carbon impacts and climate strategies.

4.0 ECONOMIC FACTORS DETERMINANT FOR VOLUNTARY DISCLOSURE

Companies that systematically monitor their carbon emissions over the life cycle of a product, while keeping an eye on the bottom line, find ways to improve the allocation of resources, eliminate waste and reduce inefficiencies; it should not be surprising that investment in reducing carbon emissions gives higher yields. The study is bases on the theoretical framework as stated in figure 1 this showing the factors the influence voluntary carbon disclosure property companies in developing countries.

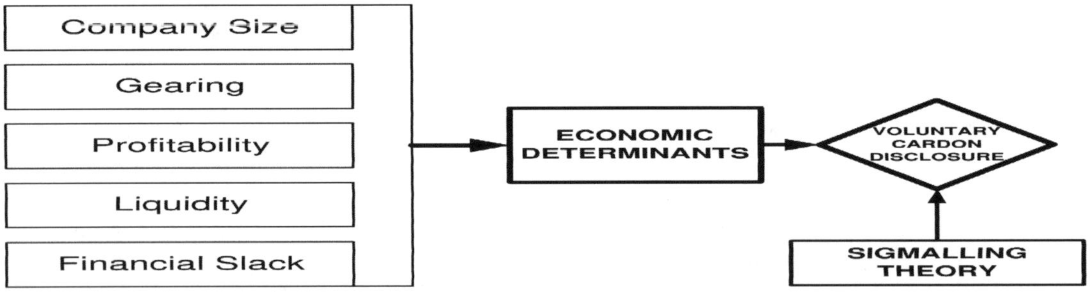

Figure 1: The Theoretical framework of the study

Company Size: The practice of social and environmental performance and the subsequent disclosure requires considerable resources in terms of funding and expertise. Large companies are considered to have more resources to cover the cost of reducing pollution and associated costs [25]. Therefore, the assumption that large companies will disclose more information than smaller firms holds. The availability of resources is of paramount importance to address the issues related to the mitigation of climate change, which often requires the company significantly operates. All studies on the disclosure of greenhouse gas emissions found a significant positive correlation between the size of the company and the disclosure of information [12, 48].

Gearing: Based on the principles of agency theory [49], designed for management of the company that is strongly concerned with gearing may disclose more information to reduce agency costs. In fact, the risk is considered by shareholders and debt holders. A company that is highly leveraged; Lenders fear that if the company was not properly managed, it is possible that there is a transfer of wealth of shareholders. In this case, if sufficient information is not provided to lenders, they will find a means, and that results to increased costs for the institutions [50]. In a study [49], the increase in spending to reduce the level of disclosure of the company, is a way to minimize costs for the agency. Findings from literature on the effect of gearing on disclosure from empirical studies are mixed. It is [25] concluded that they had no role, while another study [23] found a negative correlation. Another study [12] found no significant connection with greenhouse gas disclosure.

Profitability: [51] It is noted that profitability enables managers to a pool of resources that can be used to absorb the costs of environmental reporting. Studies argue that profitable companies are more susceptible to the public, so that interested parties may be interested in how the company makes its profit [48, 52]. Thus, profitable companies facing public pressure on how they made their profit can use information disclosure. Such disclosure of environment information justifies their profitability [53]. This disclosure may be a means of achieving public confidence and legitimacy in terms of how profits are made, not at the expenses of the environment. Empirical evidence on how greenhouse gases disclosure affects profitability is mixed. It was [23] found evidence of a significant negative correlation between profitability and disclosure with respect to GHG measure of profitability, but there is no significant relationship to the second indicator of profitability. A study [25] reported a non-significant relationship with profitability. However, other studies have found a significant positive correlation [48]. The idea seems to be consistent with the profitability of voluntary disclosure as a means of transmitting information to outside investors is seen as a tool to gain competitive advantage. The idea of this theory is that companies can use voluntary environmental disclosure to signal that their intangible assets will help to ensure future earnings. Rejection of this hypothesis would be to suggest that the financial performance or profitability due to the detection of an environmental signal connector will indicate whether it is positive or negative relationship [54].

Liquidity: Liquidity refers to the ability of the company to fulfill its short-term obligations. Based on the requirements of the theory of signals, a company with a high degree of liquidity would disclose more information in order to distinguish itself from other companies that are less liquid [55, 56]. Previous evidence has been mixed; it was [56] found that liquidity was the main factor determining a significant positive relationship between voluntary disclosures. Instead, it was [55] found that liquidity has no influence on the state.

Financial slack: The financial slack impact on the disclosure of GHG emissions, as companies allocate more financial resources to address climate change, including the disclosure of environmental information [57], financial slack has been found to enable companies to participate in new projects of which they could not participate if there were no surplus resources. It was [58] found that a company with more resources could explore new products. This affirms that more resources will influence activities such as emissions monitoring and reports and because significant investment in skills and human resources are required. It is [39] argued that the existence of unused resources allows the company to cover the costs associated with decisions on voluntary disclosure. Consequently capital expenditures in the assets and equipment may affect disclosure of greenhouse gas emissions, such as those with more advanced equipment are considered to be able to manage their emissions well compared to those with old equipment [59]. Thus, companies that make high level of capital expenditure are expected to have newer technique that will enable them to manage their emissions and, therefore, will be ready to make more disclosure. It was [60] found that companies with an aggressive environmental strategy, leading to the disclosure of relevant information on the environment often invest in new equipment and the more efficient operational environment. So invest in new, cleaner technologies and the management of being eco-friendly, management is committed to disclosing the information on this as a way to inform their stakeholders[61].

5.0 CONCLUSION

The economic factors that influence voluntary carbon disclosure in the property sector of developing countries by this review are the size of the company which determines the resource available to it. Then profitability and liquidity was determined the accessible funds of mitigation and disclosure activities. Finally financial slack influences the easy with which equipment and machinery are replaced to reduce greenhouse gas emissions.

References:

1. IPCC, *Climate Change 2007*, in *The Physical Science Basis, Contribution of Working Group I to the Fourth Assessment Report of the Intergovernmental Panel on Climate Change*, S. ThSolomon, D. Qin, M. Manning, Z. Chen, M,. Marquis, K.B. Averyt, M. Tignor and H.L. Miller, Editor. 2007, Cambridge: United Kingdom and New York, NY, USA,.
2. Hanewinkel, M., S. Hummel, and D.A. Cullmann, *Modelling and economic evaluation of forest biome shifts under climate change in Southwest Germany.* Forest ecology and management, 2010. **259**(4): p. 710-719.
3. McLeman, R. and B. Smit, *Migration as an adaptation to climate change.* Climatic change, 2006. **76**(1-2): p. 31-53.
4. Oliver-Smith, A., *Debating environmental migration: society, nature and population displacement in climate change.* Journal of International Development, 2012. **24**(8): p. 1058-1070.
5. Salazar, L.F. and C.A. Nobre, *Climate change and thresholds of biome shifts in Amazonia.* Geophysical Research Letters, 2010. **37**(17).
6. Brown, M.E. and C.C. Funk, *Food security under climate change.* 2008.
7. Chakraborty, S. and A.C. Newton, *Climate change, plant diseases and food security: an overview.* Plant Pathology, 2011. **60**(1): p. 2-14.
8. Repetto, R. and R. Easton, *Climate change and damage from extreme weather events.* Environment: Science and Policy for Sustainable Development, 2010. **52**(2): p. 22-33.
9. Bebbington, J., C. Larrinaga-González, and J.M. Moneva-Abadía, *Legitimating reputation/the reputation of legitimacy theory.* Accounting, Auditing & Accountability Journal, 2008. **21**(3): p. 371-374.
10. IPCC, *The Physical Science Basis. Contribution of Working Group I to the Fifth Assessment Report of the Intergovernmental Panel on Climate Change*, in *Climate Change 2013*, T.F. Stocker, D. Qin, G.-K. Plattner, M. Tignor, S. K. Allen, J. Boschung, A. Nauels, Y. Xia, V. Bex and P. M. Midgley, Editor. 2013, Cambridge,: United Kingdom and New York, NY, USA. h.
11. OECD, O., *Factbook 2010: economic, environmental and social statistics.* Organization for Economic Cooperation and Development, Paris, 2010.
12. Rankin, M., C. Windsor, and D. Wahyuni, *An investigation of voluntary corporate greenhouse gas emissions reporting in a market governance system: Australian evidence.* Accounting, Auditing & Accountability Journal, 2011. **24**(8): p. 1037-1070.
13. Kolk, A. and J. Pinkse, *A perspective on multinational enterprises and climate change: Learning from "an inconvenient truth" &quest.* Journal of International Business Studies, 2008. **39**(8): p. 1359-1378.
14. Pfeifer, S. and R. Sullivan, *Public policy, institutional investors and climate change: a UK case-study.* Climatic Change, 2008. **89**(3-4): p. 245-262.
15. Mol, A.P.J., *Environmental Reform in the Information Age*, in *The Contours of Informational Governance.* 2008, Cambridge.
16. Utting, P., *Regulating business via multistakeholder initiatives: A preliminary assessment.* Voluntary approaches to corporate responsibility: Readings and a resource guide, 2002: p. 61-130.
17. Levy, D.L. and P. Newell, *Multinationals in Global Governance', in S. Vachani* T.i.G.G.I.f.M.a.O. Stakeholders, Editor. 2006, Edward Elgar: London.
18. Knox-Hayes, J., *Constructing carbon market spacetime: climate change and the onset of neo-modernity.* Annals of the Association of American Geographers, 2010. **100**(4): p. 953-962.

19. Murphy, D.F. and J. Bendell, *Partners in time? Business, NGOs and sustainable development.* Discussion paper-United Nations Research Institute for Social Development, 1999(109): p. 1-71.
20. Farnworth, E. and *Climate Group: Are Investors Taking Carbon Disclosure Seriously?', Opinion*, in *www.climatechangecorp.com/content.asp?ContentID=5011*2007, Climate Change Corp: 14 November,.
21. Kolk, A., D. Levy, and J. Pinkse, *Corporate responses in an emerging climate regime: the institutionalization and commensuration of carbon disclosure.* European Accounting Review, 2008. **17**(4): p. 719-745.
22. Matsumura, E.M., R. Prakash, and S.C. Vera-Munoz. *Carbon emissions and firm value*. 2011. AAA.
23. Prado-Lorenzo, J.M., I. Gallego-Alvarez, and I.M. Garcia-Sanchez, *Stakeholder engagement and corporate social responsibility reporting: the ownership structure effect.* Corporate Social Responsibility and Environmental Management, 2009. **16**(2): p. 94-107.
24. Stanny, E. and K. Ely, *Corporate environmental disclosures about the effects of climate change.* Corporate Social Responsibility and Environmental Management, 2008. **15**(6): p. 338-348.
25. Freedman, M. and B. Jaggi, *Global warming, commitment to the Kyoto protocol, and accounting disclosures by the largest global public firms from polluting industries.* The International Journal of Accounting, 2005. **40**(3): p. 215-232.
26. Peters, G. and A. Romi, *Carbon disclosure incentives in a global setting: An empirical investigation*. 2009.
27. Reid, E.M. and M.W. Toffel, *Responding to public and private politics: Corporate disclosure of climate change strategies.* Strategic Management Journal, 2009. **30**(11): p. 1157-1178.
28. Verrecchia, R.E., *Essays on Disclosure.* Journal of Accounting and Economics, 2001: p. 32, 97-180
29. Bourgeois, L.J., *On the measurement of organizational slack.* Academy of Management review, 1981. **6**(1): p. 29-39.
30. Ullmann, A.A., *Data in search of a theory: A critical examination of the relationships among social performance, social disclosure, and economic performance of US firms.* Academy of management review, 1985. **10**(3): p. 540-557.
31. Harmeling, S., *Global climate risk index 2012: who suffer most from extreme weather events? Weather-related loss events in 2010 and 1991 to 2010. Bonn: Germanwatch;[cited 2013 Mar 5]*. 2011.
32. Deloitte, K., *Institutional study for DID–final report June 2003.* Department of Irrigation and Drainage, Malaysia, 2003.
33. Aliagha, G., QQ Q 11 Q al., *Discriminant factors of flood insurance demand for flood-hit residential properties: a case for Malaysia.* Journal of Flood Risk Management, 2015. **8**(1): p. 39-51.
34. Loi, H.K. *Flood mitigation and flood risk management in Malaysia.* in *International Workshop on Floodplain Risk Management*. 1996.
35. Liu, P.-S. and N.W. Chan, *The Malaysian flood hazard management program.* International Journal of Emergency Management, 2003. **1**(3): p. 205-214.
36. BERNAMA and *Malaysia Faces Serious Challengein Handling Floods and Drought,*, in *http://www.piam.org.my/index.php/news/,*. 2007: 21 June 2007.
37. MNREM, *Flood and Drought Management in Malaysia*, M.o.N.R.a. Environment and Malaysia, Editors. 2007.
38. Aliagha, U., et al., *Factors affecting flood insurance purchase in residential properties in Johor, Malaysia.* Natural Hazards and Earth System Science, 2014. **14**(12): p. 3297-3310.
39. Brammer, S.J. and S. Pavelin, *Corporate reputation and social performance: The importance of fit.* Journal of Management Studies, 2006. **43**(3): p. 435-455.

40. Clarkson, P.M., et al., *Revisiting the relation between environmental performance and environmental disclosure: An empirical analysis.* Accounting, Organizations and Society, 2008. **33**(4): p. 303-327.
41. Verrecchia, R.E., *Discretionary disclosure.* Journal of accounting and economics, 1983. **5**: p. 179-194.
42. Spence, M., *Job market signaling.* The quarterly journal of Economics, 1973: p. 355-374.
43. Weiss, A., *Human capital vs. signalling explanations of wages.* The Journal of Economic Perspectives, 1995: p. 133-154.
44. Morris, R.D., *Signalling, agency theory and accounting policy choice.* Accounting and business Research, 1987. **18**(69): p. 47-56.
45. Ross, S.A., *Disclosure regulation in financial markets: Implications of modern finance theory and signaling theory.* Issues in financial regulation, 1979. **5**: p. 177-202.
46. Connelly, B.L., et al., *Signaling theory: A review and assessment.* Journal of Management, 2011. **37**(1): p. 39-67.
47. Spence, M., *Signaling in retrospect and the informational structure of markets.* American Economic Review, 2002: p. 434-459.
48. Berthelot, S. and A.-M. Robert, *Climate change disclosures: An examination of Canadian oil and gas firms.* Issues in Social and Environmental Accounting, 2012. **5**(1/2): p. 106-123.
49. Jensen, M.C. and W.H. Meckling, *Theory of Firm: Managerial Behaviour Agency Costs and Ownership Structure.* Journal of Finacial Economics, 1976. **4**(4).
50. Depoers, F., *A cost benefit study of voluntary disclosure: Some empirical evidence from French listed companies.* European Accounting Review, 2000. **9**(2): p. 245-263.
51. Brammer and Pavelin, *Voluntary environmental disclosures by large UK companies.* Journal of Business Finance & Accounting, 2006. **33**(7-8): p. 1168-1188.
52. CHITHAMBO, L., *Firm characteristics and the voluntary disclosure of climate change and greenhouse gas emission information.* International Journal of Energy and Statistics, 2013. **1**(03): p. 155-169.
53. Bewley, K. and Y. Li, *Disclosure of environmental information by Canadian manufacturing companies: a voluntary disclosure perspective.* Advances in environmental accounting and management, 2000. **1**(1): p. 201-226.
54. Freedman, M. and B. Jaggi, *An analysis of the association between pollution disclosure and economic performance.* Accounting, Auditing & Accountability Journal, 1988. **1**(2): p. 43-58.
55. Aly, D., J. Simon, and K. Hussainey, *Determinants of corporate internet reporting: evidence from Egypt.* Managerial Auditing Journal, 2010. **25**(2): p. 182-202.
56. Oyeler, P., F. Laswad, and R. Fisher, *Determinants of internet financial reporting by NewZealand companies.* Journal of International Financial Management and Accounting, 2003. **14**(1): p. 26-63.
57. Kock, C.J., J. Santaló, and L. Diestre, *Corporate governance and the environment: what type of governance creates greener companies?* Journal of Management Studies, 2012. **49**(3): p. 492-514.
58. Voss, G.B., D. Sirdeshmukh, and Z.G. Voss, *The Effects of Slack Resources and Environmentalthreat on Product Exploration and Exploitation.* Academy of Management Journal, 2008. **51**(1): p. 147-164.
59. de Villiers, C., V. Naiker, and C.J. van Staden, *The effect of board characteristics on firm environmental performance.* Journal of Management, 2011: p. 0149206311411506.
60. Gao, L.S. and E. Connors, *Corporate Environmental Performance, Disclosure and Leverage: An Integrated Approach.* International Review of Accounting, Banking and Finance, 2011. **3**(2): p. 1-26.
61. Clarkson, P.M., M.B. Overell, and L. Chapple, *Environmental reporting and its relation to corporate environmental performance.* Abacus, 2011. **47**(1): p. 27-60.

Geological factors contributing to landslides: case studies of a few landslides in different regions of India

Nirmala Vasudevan[1] and Kaushik Ramanathan[2]
Amrita Vishwa Vidyapeetham (Amrita University), Amritapuri Campus, Amritapuri, Clappana P. O., Kollam - 690525, Kerala, India

E-mail: [1] nirmalav@am.amrita.edu, [2] kaushikramanathan@am.amrita.edu

Abstract. Landslides – mass movements of rock, debris or earth down a slope – are worldwide phenomena which cause significant damage and an estimated 5000 fatalities each year. They are caused by the interplay of various natural and anthropogenic factors and occur under diverse geoenvironmental conditions. In India, landslides occur primarily in the Himalayas of North India and in the Western Ghats of South India. This paper reports the results of field investigations for six landslide sites in North, Northeast and South India. We provide explanations as to why several landslides occurred at each of the sites. Our goal is to gain a deeper insight into the causes and precursors of landslides, which will facilitate more accurate identification of landslide-prone locations and enable early detection of landslide events.

1. Introduction

Landslides, mass movements of rock, debris or earth down a slope, are natural processes that have shaped much of the earth's landscape. However, with increased human settlement in unstable terrain, landslides pose a more serious threat to humans than ever before, causing an estimated 5000 fatalities each year [1].

Landslides occur when the downward forces, including the force of gravity, exceed the cohesive forces holding the landmass together, and a failure of the slope-forming material results. Several factors contribute to a landslide occurrence – geological factors such as weak, sensitive and sheared material, the presence of fissures and joints, and contrasts in permeability or stiffness of the slope forming material; morphological factors such as tectonic uplift, glacial rebound and erosion of the hill slope or toe; physical factors such as heavy rainfall, rapid snow melt and earthquakes; and anthropogenic factors such as mining, deforestation and excavation of the hill slope or toe [2].

As part of a larger study [3], we visited and conducted field investigations on several landslide sites in the Himalayas of North India; the Sikkim Himalayas; the Patkai hills of the Northeast Himalayas; and the Western Ghats of South India. The sites have varying geologies, topographies, drainage patterns and geotechnical characteristics, but have nevertheless experienced quite a few landslides. In this paper, we report the results of reconnaissance studies of the following sites (please refer figure 1):

- Bhatwari, Uttarkashi Town, Uttarakhand State, Himalayas, North India
- Chandmari, Gangtok City, Sikkim State, Himalayas, Northeast India
- Old Mangan Bazaar, Mangan Town, Sikkim State, Himalayas, Northeast India
- Ramhlun Sports Complex, Aizawl City, Mizoram State, Patkai hills, Himalayas, Northeast India

- Laipuitlang, Aizawl City, Mizoram State, Patkai hills, Himalayas, Northeast India
- Anthoniar Colony, Munnar Town, Kerala State, Western Ghats, South India

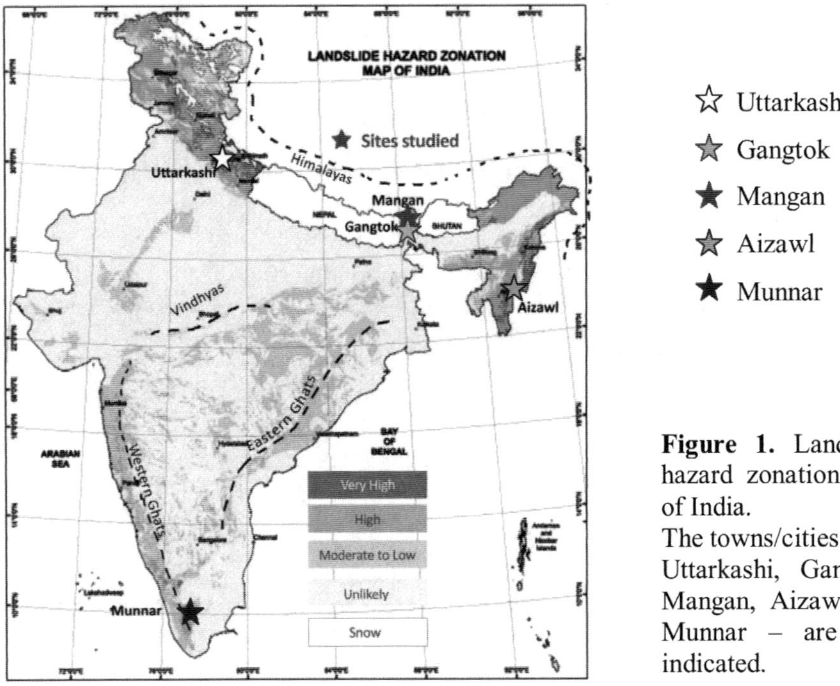

Figure 1. Landslide hazard zonation map of India.
The towns/cities – Uttarkashi, Gangtok, Mangan, Aizawl and Munnar – are also indicated.

2. Methods

We walked over each of the sites and neighbouring slides, accompanied by local geologists. Predominant rocks, minerals and geologic formations were identified. The presence of rivers, streams, and underground water sources was noted. Specific attention was paid to landslide indicators such as tilted trees, jammed doors and windows, and cracks on the ground and on structures. Old records of rainfall, land use, and landslide occurrence were consulted. The residents and local administration provided additional information. We present the results of these studies in this paper.

We have also conducted laboratory and in-situ soil tests [3], [4], and installed piezometers and inclinometers at some of the sites [4], the results of which are beyond the scope of this paper.

3. Bhatwari Landslide, Uttarkashi Town, Inner Himalayas, North India

The Bhatwari hill/village (30°49' N, 78°37' E, elevation 1610 m, figure 2), like numerous nearby hills, lies in Uttarkashi Town, along the banks of River Bhagirathi (one of the two headstreams of the Ganges river). The national highway NH 108 passes through these hills. Bhatwari, and the other hills in Uttarkashi, have experienced several landslides, particularly during the monsoons [5].

Uttarkashi is bounded by two major thrust faults – the Main Central Thrust (MCT) and the Srinagar Thrust, and is in a seismically active zone. The MCT separates two groups of rocks – the highly metamorphosed Central Crystallines (schists, gneisses, amphibolites, migmatites) that thrust southward over the less metamorphosed Garhwal Group (quartzites, epidiorites, slates). Bhatwari Hill has loose unconsolidated overburden of fluvioglacial material, while its bedrock consists of foliated mica schist sandwiched between massive jointed gneissic rocks.

Subsidence has been observed since 2009 [5]. A rainfall-triggered landslide occurred on the night of

12–13 August 2010, and another slide occurred on 16 June 2013 during the 2013 Uttarakhand floods [5]. The factors that contributed to these landslides include erosion of the hill's toe by the Bhagirathi, saturation of the loosely-consolidated overburden, the steep slope of the hill, the highly jointed bedrock, the presence of thin mica schist foliations between the jointed gneissic rocks, and the presence of water leading to the removal of fines at the overburden-rock interface.

Figure 2. Bhatwari Landslide

Figure 3. Chandmari Landslide

4. Chandmari Landslide, Gangtok City, Inner Himalayas, Northeast India

Chandmari Hill (27°20' N, 88°33' E, elevation 1459 m, figure 3) is located in the Eastern Himalayan mountain ranges of Gangtok city, the capital of Sikkim State.

The Daling Group of rocks (quartzites, phyllites, dolomites) occurs commonly, along with foliated augen gneisses having streaky character, such as the Lingtse Gneiss [6]. Bands of weathered biotite-muscovite schist are also frequently encountered. There is thick soil overburden mixed with boulders and weathered mica gneiss. The predominant minerals include quartz, feldspar and mica.

Subsidence was first observed in 1966 [7]. The landslide was fairly active during the period 1975–1976, but thereafter, it remained relatively stable until its reactivation in June 1984 as a landslide and subsidence zone [8]. A rainfall of 211 mm in 4 hours on the night of 8–9 June 1997 triggered at least nine landslides in and around Gangtok, including one at Chandmari. Rainfall-triggered landslides occurred in July 2007 and June 2011, while an earthquake on the Sikkim-Nepal border on 18 September 2011 triggered minor debris slides. Currently, downward creep is observed here.

The factors contributing to landslide occurrences include heavy rainfall, the presence of weathered rocks, the contrast in permeability in adjacent soil layers, seismic activity, and anthropogenic factors such as improper construction activities.

5. Old Mangan Bazaar Landslide, Mangan Town, Inner Himalayas, Northeast India

Mangan is the headquarters of North Sikkim District. Old Mangan Bazaar (27°30' N, 88°32' E, elevation 1137 m, figure 4), the local market place, has experienced landslides in the past [9].

Mangan is dominated by three rock types – quartz-biotite schist, biotite schist and garnetiferous mica schist. Schists are characteristically foliated and hence easily eroded. The river Teesta flows at about a kilometre away from this area and erodes the toe of the hill, leading to further instability.

The most recent landslide occurred after a rainfall of 196.6 mm on 20 September 2012. Heavy rainfall during 19–23 September 2012 triggered many more landslides in the region. Old landslides were reactivated due to heightened erosion by Raffong Khola, a tributary of River Teesta.

Figure 4. Mangan Landslide

Figure 5. Landslide at Ramhlun Sports Complex

6. Ramhlun Sports Complex Slide, Aizawl City, Patkai hills, Inner Himalayas, Northeast India
Aizawl, the capital city of Mizoram state, is set on a narrow, elongated, north-south trending anticlinal ridge, and has an average elevation of 1132 m. Ramhlun Sports Complex (23°45' N, 92°43' E, elevation 963 m, figure 5) is a densely populated locality in Aizawl City, with an escarpment to its east, and Bangla Lui, a tributary stream of Chite Lui River, flowing along its northern border.

Only sedimentary rocks are found in Mizoram. The rock formation at Ramhlun Sports Complex consists of a thin bed of fine-grained silty shale underlying a relatively thick bed of silty sandstone/siltstone, which in turn underlies well-bedded sandstone. Several joint sets, as well as fractures, are present in the rock bed. The overburden consists of sandy soil with soft shale fragments.

Creep movement has been observed since 1994. Rainfall-triggered rock falls–debris slides occurred on 19 July 2004, 14 September 2007, and during the period 10–20 August 2012. More recently, in August 2013, several houses collapsed due to land subsidence. The joints, fractures and cracks in the rock bed play a critical role in destabilising the slope. Other factors that contribute to slope instability include the abrupt escarpment to the east, the presence of pronounced bedding planes in the bedded sandstone, heavy rainfall, erosion by Bangla Lui Stream, and an improper drainage system that allows water to seep through the joints and cracks, leading to saturation of the overburden material.

7. Laipuitlang Landslide, Aizawl City, Patkai hills, Inner Himalayas, Northeast India
Laipuitlang Hill (23°44' N, 92°43' E, elevation 1134 m, figure 6) is another densely populated locality in Aizawl City. The bedrock consists of shale underlying bedded sandstone, while the overburden consists of sandy soil interspersed with fragments of shale. Joints are present in the rock bed.

The area was once a sandstone quarry; quarrying destabilised the slope, and rockslides occurred in 1957 and 1968. Quarrying was then discontinued. Further rockslides occurred on 5 May 2002 and 25 November 2007 following indiscriminate excavation of the hill. [10], [11]

The next sign of landslide activity was in 2010 when cracks developed in a five-storied building. In September 2012, cracks appeared in the rock bed and on the walls of a large Public Works Department (PWD) building that was located on the crown of the hill. The base of the building was on the sandstone-shale interface. In the interest of safety, geologists from the Directorate of Geology and Mineral Resources, Government of Mizoram, strongly recommended the demolition of the building, but timely action was not taken. On account of improper drainage, rainwater penetrated the foundation of the PWD building and further weakened it. At 2:30–3:00 am, 11 May 2013, after a week of heavy rainfall, the building and its foundation slid downhill burying ten houses and a church. The top sandstone layer was almost completely eroded – only a little remains along the left flank. [10], [11]

Anthropogenic factors – quarrying, indiscriminate excavation of the hill, the construction of the large PWD building on the sandstone-shale interface – played a major and obvious role in the Laipuitlang landslides. Other causative factors include the presence of vertical joints, heavy rainfall, improper drainage, and water seepage through rock joints and bedding intersections. Shale can be a problematic material; water caused the weathering of the shale foundation, and this weakening and the subsequent sliding of the foundation played a pivotal role in the 2013 landslide occurrence.

Figure 6. Laipuitlang Landslide

Figure 7. Anthoniar Colony Landslide

8. Anthoniar Colony Landslide, Munnar Town, Western Ghats, Southwest India

Anthoniar Colony (48°08' N, 11°31' E, elevation 1520 m, figure 7) is located in the tourist town of Munnar in Kerala State.

Precambrian crystallines consisting of hornblende-biotite gneiss, biotite gneiss, granite gneiss, charnockite and pink granites form the main rock types at Munnar [12]. The bedrock at Anthoniar Colony consists of granite gneiss, underlying thick overburden of lateritic soil. Predominant minerals include kaolinite and illite; quartz sericite bands are also seen. Fissures are observed on the hill [4].

Two slides have occurred at Anthoniar Colony. The first one in 1926 was most likely caused by the combined action of rainfall and weak geology, since the place was uninhabited then. More recently, a torrential downpour of 451 mm on 25 July 2005 triggered landslides here and at two other sites in Munnar. This slide was exacerbated by human activity such as the excavation of the hill [4].

9. Results and Discussion

1. We have presented the results of our field investigations – the geology of the landslide sites, and the causes of the slides – in the preceding sections. India consistently has one of the highest incidences of landslides, but there is insufficient information on the geology of the sites and the slides themselves. A record and subsequent study of such data will facilitate better understanding of landslide causes and consequently, a more accurate prediction of landslide events.

   Our findings on the Bhatwari slide (section 3) can be extrapolated to other landslide events in the vicinity. We visited the nearby slides and noted that the hills all lie on the banks of the Bhagirathi, have similar geology and slope angle as Bhatwari Hill, and like the latter, they too experienced landslides on 16 June 2013. We are led to believe that as in the case of Bhatwari Hill, these landslides were also caused by the combined action of toe erosion by the Bhagirathi, and saturation of loosely-consolidated overburden induced by excessive rainfall.

   Caution should be exercised in extrapolating results. Both Ramhlun (section 6) and Laipuitlang (section 7) in Aizawl have similar geology, but the slides at Laipuitlang were caused primarily due to anthropogenic factors, while geological factors played a predominant role in the Ramhlun slides.

2. Our results illustrate that landslides occur under diverse geoenvironmental conditions (sections 3 – 8). Landslides are seldom caused by a single factor; rather, they usually occur due to the combined effect of several causes. These causes may increase shear stress, or decrease material strength, or simultaneously increase shear stress and decrease material strength [2]. For example, erosion at Bhatwari, Mangan and Ramhlun Sports Complex leads to a removal of lateral support, which causes an increase in shear stresses acting on the hill. Another example is the presence of joints, foliations, bedding surfaces, faults, fissures or discontinuities which are seen at the sites described in this paper and which contribute to the low strength of the slope-forming material. Yet another example is rainwater infiltration which increases soil weight (and thus, shear stress) and simultaneously decreases the strength of the slope-forming material.

3. Landslides occur either due to natural causes alone as in the case of the 1926 Anthoniar Colony slide, or due to a combination of natural and anthropogenic causes as seen in the Laipuitlang slides.

10. Conclusion

We have presented an account of the geology and landslide causes for six landslide sites in different regions of India. The causative factors have been examined in detail.

References

[1] Petley D N, Dunning S A and Rosser N J 2005 The analysis of global landslide risk through the creation of a database of worldwide landslide fatalities *Landslide Risk Management* (Amsterdam: Balkema) pp 367–74

[2] Cruden D M and Varnes D J 1996 Landslide types and processes *Landslides: Investigation and Mitigation* (*Transportation Research Board Special Report* 247) ed A K Turner and R L Schuster (Washington DC: National Academy Press) chapter 3 pp 67–71

[3] Vasudevan N and Ramanathan K 2015 Geotechnical characterization of a few landslide-prone sites in India *Proc. 6th Int. Geotechnical Symp. on Disaster Mitigation in Special Geoenviron. Cond.* (Chennai: Indian Geotechnical Society, Chennai Chapter) pp 509–12

[4] Ramesh M and Vasudevan N 2012 The deployment of deep-earth sensor probes for landslide detection *Landslides* **9(4)** pp 457–474

[5] Dangwal D P, Chauhan N, Ghosh M and Ghosh T 2014 *Preliminary Slope Stability Assessment of the Recent Disaster Affected Areas of Uttarkashi District, Uttarakhand*, (Kolkata: Geological Survey of India)

[6] Paul D K, Mcnaughton N J, Chattopadhyay S and Ray K K 1996 Geochronology and Geochemistry of the Lingtse Gneiss, Darjeeling - Sikkim Himalaya: Revisited *J. Geol. Soc. India* **48(5)** pp 497–506

[7] Rawat R K 2005 Geotechnical investigations of Chandmari landslide located on Gangtok-Nathula road, Sikkim Himalaya, India. *Him. Geol., Dehra Dun* **26(2)** pp 309–22

[8] Basu S R and De S K 2003 Causes and consequences of landslides in the Darjeeling-Sikkim Himalayas, India. *Geographia Polonica* **76(2)** pp 37–52

[9] Larsen J O, Grimstad E, Bhasin R, Dhawan A K, Singh R and Verma S K 2000 Landslides and their mitigation in Gangtok, Sikkim *J. Nepal Geo. Soc.* pp. 585-590

[10] Verma R 2014 Landslide Hazard in Mizoram: Case Study of Laipuitlang Landslide, Aizawl. *Int. J. of Sci. and Res.* **3(6)** pp 2262-6

[11] Laldinpuia, Kumar S and Singh T N 2014 11th May, 2013 Laipuitlang Rockslide, Aizawl, Mizoram, North-East India. *Landslide Science for a Safer Geoenvironment* (Springer International Publishing) pp 401–5

[12] Abraham P B and Shaji E 2013 Landslide hazard zonation in and around Thodupuzha-Idukki-Munnar road, Idukki district, Kerala: A geospatial approach *J. Geol. Soc. India* **82(6)** p 649

Market Potential Study for Standing Cabin Concept for Domestic Low-Cost Commercial Airlines in Malaysia

Fairuz I. Romli, Norhafizah Dasuki and Mohammad Yazdi Harmin

Department of Aerospace Engineering, Faculty of Engineering,
Universiti Putra Malaysia, 43400 Serdang, Selangor, Malaysia

fairuz_ir@upm.edu.my

Abstract. An affordable air transportation has become the operational aim of many airlines these days. This is to cater the growing air travel demands from people of different social and economic status. One of the revolutionary proposals to reduce the operational costs, hence the flight ticket price, is by introducing the so-called standing cabin concept. This concept involves transporting passengers during the entire flight in their standing position with a proper support of a vertical seat. As can be expected with many new inventions, despite its clear advantages, the concept has been met with mixed reactions from the public. This study intends to establish whether the standing cabin concept has a market potential to be implemented for domestic flights in Malaysia. The public perception is determined from collected data through a survey done at two major local low-cost airport terminals. It can be concluded from the results that the concept has a good market potential for application on flights with duration of less than two hours.

1. Introduction

Today, there is a significant rise in the need for people with different social and economic background to travel from one point to another in a shorter time duration, sometimes on a daily basis. Due to this situation, air transportation has transformed itself into an essential travel means and no longer caters only for luxury trips like it mostly did in the past. Accordingly, to cater for passengers market that is now extended across different social and economic classes, the operational paradigm of many airlines is shifting towards providing more affordable flight services to their customers. This means having a cheaper flight operation to enable competitive reduction in their offered ticket prices. This changing operational paradigm, and its significant success, is evident by the large numbers of profitable low-cost carriers that are serving the worldwide market these days.

As the air transportation industry continues to grow, the market competition is also becoming more competitive. For most low-cost airlines, to remain operationally profitable, their operation has a high break-even load factor that could be lifted even higher by market factors such as increasing fuel prices [1]. This situation leads them to look for alternative ways to keep their low operational costs in order to still be able to offer low ticket fares to their passengers. A low-cost airline in China, Spring Airlines has started to look at the possibility of having standing-only passenger cabin on their aircraft fleet back in 2009. It is projected that such cabin design can increase the passengers cabin capacity by about 40% than that of the conventional cabin design and reduce the cost by as much as 20% [2]. Later in 2012, a major low-cost airline in Europe, Ryanair has further picked up on the idea and conducted a series of test flights with modified section of passenger cabin on their aircraft as standing-only section [3]. This development can be taken as a good indication that the standing cabin idea is growing in acceptance by the public and airlines, though there are still several issues that need to be addressed and resolved before the concept can be implemented in commercial flights. In terms of regulations, standing cabin concept is principally legal by current standards of several governing aviation authorities. The Federal Aviation Authority (FAA) does not specify that passengers have to be in their sitting position during takeoff and landing while the Air Transport Association (ATA) does not enforce any specific standard for seat comfort or seating configuration onboard the cabin [4]. According to current aviation rules, it is adequate if it can be proven that passengers have been properly secured by a certified mechanism, even if they are standing during takeoffs and landings.

The concept is highly suitable for the Malaysia domestic air transportation market because most of its flights are within two hours duration, which has been suggested as a good duration for the standing cabin concept application [5]. This limitation of flight duration is mainly proposed to avoid any health issues or fatigue to passengers. Similar to other worldwide markets today, domestic flights in Malaysia are progressively dominated by the local low-cost carrier, AirAsia. In just four years after its market introduction, AirAsia has managed to secure 30% market shares of the local domestic market [6]. The low-cost operational paradigm of AirAsia can be seen to revolutionize the domestic air transportation sector in Malaysia, opening up the market to many new customer segments who never use flights for domestic travel due to comparatively high ticket prices prior to its service entry. This can be seen from the increase in recorded number of flying passengers after AirAsia came into service, which highlights high demands for low-cost air transportation services in Malaysia. However, despite the considerably low ticket fare, there are many people in Malaysia who still choose cheaper domestic transportation options such as buses and trains. Hence there is still a potential market to be explored if the airlines can further reduce their operational cost and subsequently their offered flight ticket prices to become more competitive against other transportation options. A study conducted for the domestic Malaysian flight markets indicates that the flight ticket prices can be reduced by as much as 26% when using the standing cabin concept [7]. Further discussion on how on the standing cabin implementation can help improve the competitiveness of low-cost airlines can be found in Ref. [8]. This can be the competitive solution for the airlines to make domestic air transportation more affordable for people in Malaysia.

Acknowledging that the success of future standing cabin implementation also heavily depends on the public acceptance of the concept, the objective of this study is to analyze the public perception and establish whether there is a market potential for the concept to be implemented.

2. Standing Cabin Concept for Domestic Flights

Standing cabin is a concept of passenger cabin design arrangement where the passengers are primarily transported in their standing position throughout the flight. A graphical depiction of this cabin concept, which is considered for this study is illustrated in Figure 1. In addition to enabling the airlines to fit in more passengers onboard their aircraft fleet, it is also expected that the vertical seats used in the cabin concept are thinner and lighter than the current cabin seats. This helps to improve flight performance and also operational costs in the long run through potential savings of fuel.

The primary element of the standing cabin concept is the vertical passenger seat design. To date, no full-standing vertical passenger seat design has been approved for use in commercial transport aircraft. The closest available seat design that resembles full-standing seat considered in this study is the one proposed by Aviointeriors Company, one of the leading aircraft seat and interiors manufacturers in the world. The seat is called SkyRider, which is designed as ultra-high density seat to allow the possibility of reducing ticket prices while maintaining sound profits for low-cost airlines. It has been advertised that, even with reduced seat pitch, adequate passengers' comfort level can be expected with SkyRider since the seating position is just like "riding a tourist motor-scooter". This seat design was unveiled at Aircraft Interiors Expo Americas 2010 in Long Beach, California and back then, it was said to be in its final testing stage.

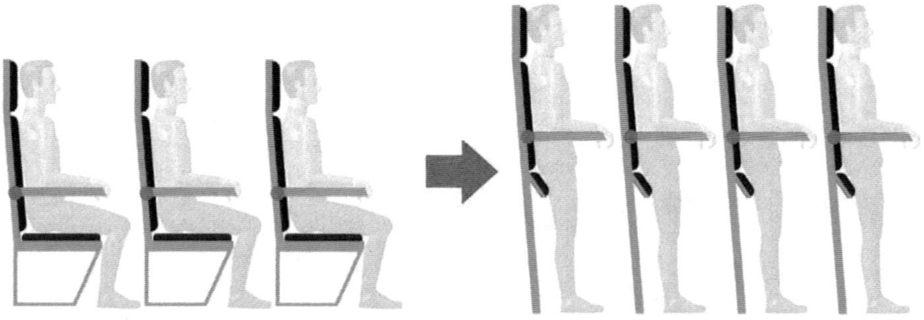

Figure 1: Transformation of Conventional Passengers Cabin into Full-Standing Passengers Cabin

The idea of vertical passenger seat for aircraft cabin has been around since back in 2006, which has been visualized to be comprised of a vertical bench with shoulder harnesses and arm rests [9]. Airbus, one of the world's leading aircraft manufacturers, have seriously considered the idea for their aircraft at one point and patterned their proposed design concept [10]. Unfortunately, not much detail has been available regarding the design of that vertical seat. In fact, no published formal study on a vertical seat design has been found in public domain during this study apart from the authors' work in Ref. [5] and Ref. [11]. The lack of published studies in public domain regarding the feasibility and viability of its implementation into commercial aircraft design makes its realization to become overlooked.

Apart from safety concern, another big issue that seems to bother many potential passengers with this standing cabin concept is its perceived low comfort level, mainly because they are expected to be standing instead of sitting throughout the entire flight duration. An ergonomics study on the standing cabin concept, however, has argued that such standing posture will not notably pose higher health risk to the passengers in comparison with that of current sitting posture for short range flights [5]. In fact, the standing posture might even be more comfortable for passengers than current seating arrangement, especially when many airlines continue to reduce the seat's pitch and width to the minimum regulated measures in order to increase their cabin capacity.

3. Data Analysis and Discussions

The survey was conducted at two major low-cost airport terminals in Malaysia: SkyPark Terminal at Sultan Abdul Aziz Shah Airport and Kuala Lumpur International Airport 2 (KLIA 2). Both terminals are primarily used as the main hub by local low-cost airlines for domestic flight operation in Malaysia. The study has been conducted both in weekends and weekdays, and in peak hours and off-peak hours. The peak hours here is assumed to be between 10am to 6pm, when the departures and arrivals of most flights are scheduled at the terminals. The target participants for this survey are people who frequently use flights as their means of travel. All in all, 1000 respondents have been surveyed for this study and the results are tabulated in Table 1.

In the survey questionnaire, respondents are asked to answer questions regarding several issues or concerns that have been previously established for the implementation of standing cabin on domestic flight operations in Malaysia. They are also asked to give their opinions and suggestions for the design of the standing cabin. For this paper, the collected survey data is analyzed to provide some insights on the market potential of the standing cabin concept. Table 1 gives a simple distributional illustration of respondents based on their response to the implementation of standing cabin concept, either positive or negative. It can be seen that about 47% of the surveyed respondents stated their willingness to try the standing cabin concept provided that the flight ticket price will be reduced. In contrast, about 53% of the respondents are not in favor with the implementation of the standing cabin, even with lower fares. This shows a clearly divided marketability of the cabin design concept but nonetheless, it highlights a possibility that the idea can still be marketed but need to be focused to certain categories of travelers.

Table 1: Statistics of Survey Data Results

If we introduce a standing cabin in the aircraft and reduce flight ticket price, would you consider to travel by the flight?		**YES (47.3%)**			**NO (52.7%)**		
Average age of respondents		38.9 years			43.9 years		
Gender of respondents	**Male**	51.1 %			48.9 %		
	Female	41.7 %			58.3 %		
Average annual income of respondents		RM 10,120.43			RM 21,789.81		
Average annual travel frequency		64 times			72 times		
Common of luggage carrying mode		Carry-on 72%	Cargo 21%	Both 7%	Carry-on 63%	Cargo 27%	Both 10%
Ability to stand within two hours		Yes 74%	No 26%		Yes 0%	No 100%	

From Table 1, it can be observed that the average age of respondents who do not favor the standing cabin is higher than those who are in favor. This is in line with the initial expectation that the concept will appeal more towards the younger travelers. In fact, the mode of age of the supportive respondents is recorded between 23 to 25 years old, indicating a strong inclination towards young travelers. The overall population of the surveyed respondents is about 60% males and 40% females. The results show that, while male respondents are rather even with a slight bias towards favoring the new cabin concept, the female respondents are notably skewed towards non-supportive responses. This is rather expected due to extra physical nature of having to stand inside the cabin as imposed on the passengers by the new concept instead of sitting in current conventional cabin. Furthermore, a big difference is evident in average income of respondents between the two response categories, whereby those who positively welcome the new concept seem to be earning much less than those who are non-supportive of the idea. This highlights that the new concept appeals more towards the travel need for passengers in the lower social and economic classes, who are in fact the majority customers for low-cost airlines in domestic flights in Malaysia.

Among those who support the idea of the standing cabin, majority of them carries their travel items as hand luggage. This fits well into the operational notion of the standing cabin concept, where further operational weight and cost savings can be achieved with a lower cargo payload from the passengers. An overwhelming 72% of travelers who favor the standing cabin commonly travels with only carry-on luggage. Another positive remarks that can be taken from the survey data is the high number of travel frequency of the respondents who in favor of the concept. Close to 90% of people in these group of positive respondents stated that they will consider to choose aircraft as their travel means if the ticket price is similar or just slightly more expensive than the price of other alternative public transportation such as buses or trains. On average, they commonly travel about 64 times per year, indicating the big market potential to be tapped into with the standing cabin concept. Lastly, it should be noted that the standing cabin concept has been proposed to work well for flights within two hours duration due to fatigue concerns that may affect the comfortability of the passengers. As indicated by the survey data results, an overwhelming 74% of the positive respondents said that standing for two hours is okay for them, further supporting the feasibility of the new cabin concept.

4. Conclusion

In general, the distribution of the responses reflects a divided market support for the implementation of the standing cabin concept for domestic flights in Malaysia. It can nevertheless be concluded that the market potential for this new concept does exist and should be capitalized by the airlines. The support for the standing cabin mostly comes from respondents who fit the characteristics of low-cost airlines' target passengers. As can be summarized from analysis of data presented in Table 1, target passengers for the standing cabin concept are frequent travelers who are mostly males, of age below 40 years old, earning in the range of middle economic class or below, and often travel with just carry-on luggage.

Due to a clearly divided acceptance of the concept based on the survey results, the standing cabin is suggested to be implemented as an alternative seating class below current conventional first, business and/or economy classes. This is mainly because it can be expected to be hard for airlines to fill up the entire aircraft cabin with only passengers who are willing to take the standing cabin option. It should be noted that the standing cabin concept will have some operational limitations in terms of the suitable types of passengers for its working mode. For instance, pregnant women, people with health issues or physical disabilities, infants and small children are not suitable to fly using the standing cabin concept. If the cabin is designed only for standing passengers, the airlines may lose more than half of its current passengers according to the projection of the survey result. Hence it is better to give the passengers an option for them to choose which seating class suits their travel preference and need at that moment.

All in all, it can be concluded that the market potential exists for the application of standing cabin concept for domestic flights in Malaysia but further research study needs to be done to ensure that its implementation can provide the utmost benefits to the airlines and also to the passengers. The way that the concept is being implemented on the aircraft cabin and the flight routes to be served with it will be instrumental in securing its market success.

Acknowledgements

The authors acknowledge the financial support for this research study from Universiti Putra Malaysia, Malaysia through research grant GP-IBT/2013/9407100. Moreover, we like to thank the management of Kuala Lumpur International Airport 2 (KLIA 2) and Skypark Terminal at Sultan Abdul Aziz Shah Airport for allowing us to conduct the public survey at their facilities. Lastly, our appreciation goes to Norashikin Othman, Hassan Rashid, Mohammad Rijaluddin Bahiki, Muhammad Faizzuddin Ismail, Khairul Afiq Rahim and Ruhana Omar for their help in conducting the survey.

References

[1] G. Dunn, "Low-Cost Carriers: Growth Expectations", http://www.flightglobal.com [Accessed: January 2014]
[2] P. Pae, "Chinese airline may offer cheaper fares to passengers who stand during short flights," Los Angeles Times, vol. 4, July 2009
[3] S. Kerridge, "Ryanair – Approval for Standing Only Cabin Areas", http://blog.france-airport-guide.com/2012/04/ ryanair-approval-for-standing-only.html [Accessed: December 2013]
[4] C. Elliot, "One Day, That Economy Ticket May Buy You a Place to Stand", http://www.ny-times.com/2006/04/25/business/25seats. html [Accessed: December 2013]
[5] F. I. Romli, A. N. Asmadi and N. Dasuki, "Ergonomics Study of Vertical Seat Design for Standing Cabin Concept in Commercial Transport Aircraft", International Review of Aerospace Engineering, vol 8, no. 3, 2015
[6] O. J. F. Connell and G. Williams, "Passengers' Perceptions of Low Cost Airlines and Full Service Carriers: A Case Study involving Ryanair, Aer Lingus, Air Asia and Malaysia Airlines", Journal of Air Transport Management, vol. 11, no. 4, 2005
[7] F. I. Romli and N. Dasuki, "Preliminary Study of Standing Cabin Concept for Domestic Commercial Flights in Malaysia", Proceeding of 2nd International Conference on Advances in Mechanical and Aeronautical Engineering, Bangkok, Thailand, 2013
[8] F. I. Romli, A. R. Mohammad Noor and N. Dasuki, "A Study on Potential Standing Cabin

Effects in Improving the Competitiveness of Low-Cost Airlines", International Journal of Engineering and Technology, vol. 6, 2014

[9] F. Gillete, "Ryanair's O'Leary Mulls One-Euro Toilets, Standing Passengers", Bloomberg, 3 September 2010

[10] J. G. Angerani, A. Baatz, S. Hiesener, H. Merensky, Airbus Deutschland GmbH, Hamburg, Patent. No. US 6,467,728 B2, Date of Patent: 22 October 2002

[11] A. R. Mohamad Noor, F. I. Romli, "Conceptual Design of Vertical Passenger Seat for Standing Cabin in Commercial Transport Aircraft", Journal of Advanced Management Science, vol. 1, 2013

IOP Conf. Series: Earth and Environmental Science **30** (2016) 012013 doi:10.1088/1755-1315/30/1/012013

Usability Evaluation Survey for Identifying Design Issues in Civil Flight Deck

Negin Ozve Aminian, Fairuz Izzuddin Romli and Surjatin Wiriadidjaja

Department of Aerospace Engineering, Faculty of Engineering,
Universiti Putra Malaysia, 43400 Serdang, Selangor, Malaysia

negin_oa@yahoo.com

Abstract. Ergonomics assessment for cockpit in civil aircraft is important as the pilots spend most of their time during flight on the seating posture imposed by its design. The improper seat design can cause discomfort and pain, which will disturb the pilot's concentration in flight. From a conducted survey, it is found that there are some issues regarding the current cockpit design. This study aims to highlight potential mismatches between the current cockpit design and the ergonomic design recommendations for anthropometric dimensions and seat design, which could be the roots of the problems faced by the pilots in the cockpit.

1. Introduction

Flight deck is the place where the pilots control the aircraft during different flight phases. Since flight safety is of paramount importance, good physical and psychological conditions of the pilots are critical issues that have to be maintained by means of, among other things, proper design of cockpit's facilities and navigational instruments. An aircraft is a large class of systems that require active interaction or continuous control by the pilot, thus the cockpit design must address the inherent sensing and reaction capabilities of pilots as a critical constraint. This is in addition to a range of associated human-machine interfaces such as design and placement of controls, displays and seat position.

Human factors and ergonomics, which has also been known as comfort, functional or user-friendly system design, is the practice of designing products, systems or processes that takes proper account on the interaction between them and the people who are intended to use them. This particular field has been developed significantly through contributions from many other disciplines. Basically, designing products or systems that are concerned with human factors and ergonomics follow two main goals: the health and also the productivity of the people who use them. A good ergonomic design is concerned with the 'fit' between user and their technological tools [1]. With regards to the cockpit design, a good interface design enables the pilots to easily access all the required control panels in any situations.

The feeling of discomfort and low back pain complaints during flights have been reported by pilots around the world each year. During a long flight or flying in improper physical conditions, a pilot may feel easily fatigue and stressed due to many possible reasons [4]. In fact, most pilots have accepted this as challenges in their working environment. However, a good cockpit design, especially for pilots who have to effectively sit in such a cramped space for prolonged time duration, should remove discomfort and keep their muscles in relaxation mood, instead of developing a habit among pilots to endure the pain and/or ignored it [2]. Though there have been a range of pilot's seat designs developed based on standards and some changes have also been made in the seat's materials, dimensions and hardness to provide comfort, the complaints still remain. Based on a conducted survey on five different transport aircraft designs, it has been found that there are mismatches between the standards and anthropometric analysis of pilot's seat for modern civil aircraft [3]. With this perception in mind, this paper intends to report on some of the potential mismatches and give some recommendations to alleviate the situation.

2. Methodology

A questionnaire consisting of 90 questions is developed and distributed to respondent pilots, which is aimed to obtain their comments and feedback regarding the ergonomic problems of the cockpit design based on their experiences. Note that for this study, the main focus is placed on the pilot's seat design. The respondents are asked to rate the issue or parameter in question using a numbering scale between 0 and 20, which is adapted from Ref. [5]. The scale reflects their satisfaction on various parameters of the seat, where the participant may select a number between two extreme conditions, for example, either too low (20) or too high (0), while the number ten (10) is considered as the correct design value. Based on their responses to the questions, the answers are collected and averaged. These information may be used to identify as to whether a real problem exists in the cockpit's design.

Overall, 20 pilots of different nationalities, who are working international flight routes for different airlines, have responded to the survey. 95% of these respondents are males while the remaining 5% are females. More information on the demographic of the respondents on the type of aircraft that they are operating is depicted in Figure 1. This background information is vital to be able to analyze whether the responses are biased towards certain aircraft designs or not.

The summary of results from the survey, in terms of the pilot's seat comfort, is shown in Figure 2. As can be observed, it shows that the current existing pilot's seats do not offer the opportunity to pilots to establish a comfortable sitting posture. In this case, the anthropometric dimensions might have been selected unsatisfactorily or improperly considered. The results can be used to indicate which part of the seat requires design improvements and the seat design parameters are compared with standards of anthropometric dimensions to find the mismatch with the current seat design.

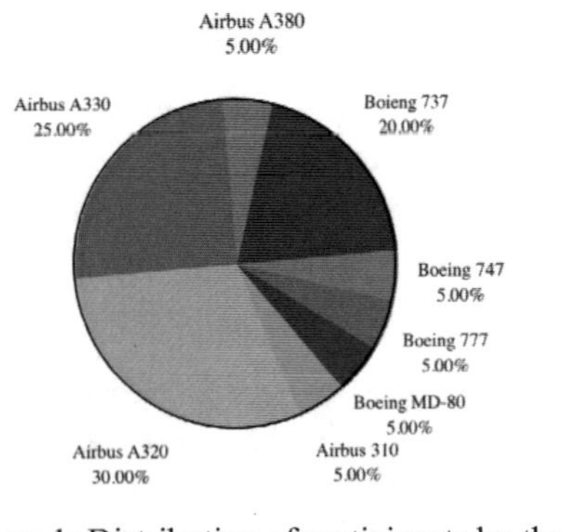

Figure 1: Distribution of participants by the types of aircraft they operate

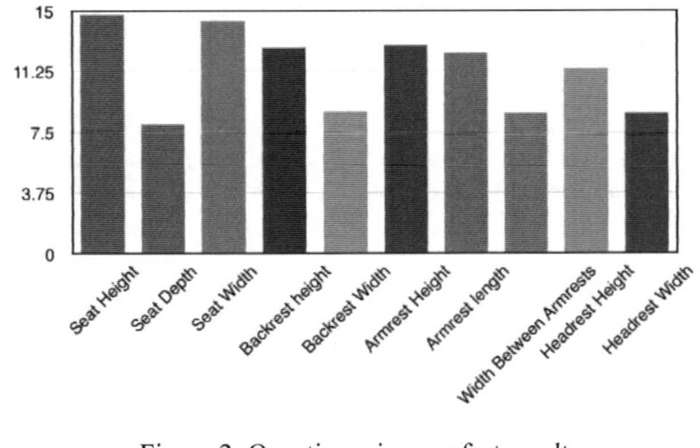

Figure 2: Questionnaire comfort result

3. Data Analysis

According to the Webster's Dictionary, comfort is defined as a feeling of relief and enjoyment. Some experts define comfort as a state of being pleasant between human and its environment, and situation. Others simply explain comfort as the absence of discomfort. It is common that pilots on international flights may fly up to about 10 hours continuously. After a few flight hours, pilots usually need to stand and walk to relieve pressure on their back and legs due to the long sitting. The usual and spontaneous pilot's body movements during sitting in the cockpit is by changing from one buttock to the other, and after a while, pushing himself forward but the shoulder harness holds him back. Hence he starts to feel aching under one of his thighs. In a little while, the pain increases and spreads throughout both thighs. When the pilot or the air crew suffers unbearable pain, they have challenges in their thoughts that may lead to losing their concentration on flight controls and focus on the body pain [6]. On the other hand, fatigue is caused by sleepiness and tiredness. A main symptom of fatigue is the feeling of exhaustion. The most common types of fatigue between pilots are visual, monotony, chronic, and circadian which extremely lead to reduce pilot's reaction time and make serious errors [7]. In order to perform safely, the pilots have to be very comfortable in the flight deck. Unfortunately, the pilots have not been taught methods to be comfortable and how to increase it to optimize their limited work station even in seating posture for long hours in the air [7]. It might be hard to assume that a lack of anthropometric design consideration may lead to tragic accident when the flight deck systems have been improperly designed for the specific population.

Anthropometric data represents both static and dynamic dimensions like shoulder breadth and grip strength, respectively. One of well-known anthropometric databases is the US Army Anthropometric Survey (ANSUR) that was collected on 1988 by the US military. In addition, another comprehensive database is the Civilian American and European Surface Anthropometric Resource (CAESAR), which is collected from US and Europe population [8]. CAESAR represents data for both males and females from few different regions: North America, Italy and Netherlands [9]. Another important ergonomics reference in this field is called "Body Space" written by Pheasant, which is also a collected data from both North America and Europe population [10]. Table 1 shows the anthropometric data measurement obtained from different literatures and ergonomics experts for the normal office chair design.

The anthropometric dimension applied for the overall seat design is set corresponding to the 5th of female to 95th of male as the overall population for the minimum and maximum design respectively. It means that it is compatible with 90 percent of users [12]. The seat should accommodate a wide range of population, and values between these percentiles is sufficient for most users. However, this is not meaningful that a woman with 5th percentile of sitting height has thigh length shorter than 5th percentile, therefore the design is performed based on her vision height and she may feel pressure while sitting for long hours on the back of her knees [13].

Table 1: Anthropometric data for office chair design [5, 11]

Seat Requirements	Anthropometric Measurement	BSR/ HFES100.2002	BIFMA	DREYFUSS	WOODSON	Grandjean	Diffrient
Seat height	Popliteal Height+ Shoe Measure	38.36	38.10 – 50.50	36.80-48.30	38.10-45.70	38-53	35-52
Seat depth	Buttock Popliteal Length - Clearance Measure	More than 43	49.90	40.60	40.60	38-42	33-41
Seat width	Hip Breadth in sitting posture+ Cloth Measure	Less than 45	More than 45.70	40.60 -55.90	48.30	40-45	41
Backrest lumbar	--------	12-15	15-24.90	17.80-29.20	17.80-25.40	NA	NA
Seatback width		more than 31.5	NA	NA	NA	32-36	33
Backrest height (upper body support)	--------	Less than 45	More than 31	More than 33	More than 20.30	48-50	NA
Armrest height	Elbow Rest Height	18-27 (adjustable)	17.50-27.40	19.10-25.40	21.60	NA	18-25
Armrest length	-------	NA	NA	25.40-30.50	30.50	NA	15-21
Distance between armrests	Hip Breadth in sitting posture + Cloth Measure	46	45.70	More than 48.30	48.30	NA	48-56

*Dimensions are in cm.

3.1. Seat Height
The 5th percentile of female to 95 percentile of male popliteal height should be considered in the seat height dimension. Based on previous literatures, 2.5 cm should be added as shoe heel allowance [11].

3.2. Seat Depth
According to ANSUR anthropometric dimensions, 95th percentile male buttock popliteal length is 54 cm, which is 10 cm greater than the 5th percentile female length. On the other hand, in CEASAR, 95th percentile male buttock popliteal length is 48 cm (12 cm greater than the 5th percentile female length). Therefore, a seat cushion length increase of 10 cm should be considered as the maximum dimension, especially for pilots who are more men than women [13].

3.3. Seat Width
For the dimension of the seat width, the widest hip breadth among the sample population should be measured and applied. In seated position, woman have larger hip breadth than male anthropometric dimensions [13]. The seat width should be comfortable not only for pilot's seat and cloth allowance but also be comfortable for their arms. It has been recommended 95th percentile of hip breadth with considering 4 to 5 cm for cloth allowance [14]. Another recommendation is to use 95th or 99th of hip breadth female and added 5 cm for heavy clothes [15].

3.4. Backrest Height
For dimension of the upper body support part of the backrest, the sitting shoulder height is considered. The 5th of female to 95th percentile of male is applied [16].

3.5. Armrest Height
The 5th percentile female to 95 percentile male of sitting elbow height data is a determinant for the anthropometric dimension in the armrest height. For the fixed armrest, only 5th percentile of female is considered. However, for this case, the armrest should be adjustable to be more compatible with wide international population. 5th to 95th seated elbow height is applied [17].

3.6. Armrest Length
Armrest are useful for upper limps support and for keeping body balanced. The best anthropometric measurement for armrest length is the elbow-to-wrist length of 95th percentile of male. It should be supported at least two thirds of forearm to be more comfortable [17].

3.7. Width between Armrest
Armrests should accommodate 95th percentile female that have the largest hip breadth in order to be comfortable, with enough clearance for clothes [18]. Based on other ergonomists, they believe that armrests clearance should be able to accommodate the 95th percentile elbow breadth male for the minimum separation between armrests [15]. Other experts apply 5th female to 95 male elbow breadth with additional three centimeters to account for heavy clothes [19].

3.8. Backrest Width
The dimension of backrest width should be determined using the shoulder width or bideltoid breadth [20]. Reynolds recommended 95th percentile chest width of male to determine the minimum width in the upper part of the backrest seat while Maertens recommended 95th percentile waist breadth of male to specify the backrest width [13].

3.9. Headrest Height
For the headrest height, dimensions from chin to crown needs to be measured. To accommodate more population, 5th percentile of female to 95th percentile male for minimum and maximum headrest is applied [21].

3.10. Headrest Width
For the headrest width, the maximum breadth across the ears is measured. To accommodate with more population, 5th percentile of female to 95th percentile male for minimum and maximum headrest is applied [21].

All in all, the comparison between AS290B (Aerospace standards and biomechanical requirements) and two anthropometric database standards: ANSUR and CAESAR is summarized in Table 2.

Table 2: Anthropometric data for the flight deck seat design [3, 22, 23]

Seat Requirements	AS290B	ANSUR	CAESAR
Seat height	33 - 51	38 - 51.5	33.44 - 44.87
Seat depth	41 - 45	44 - 54.60	36.66 - 48.09
Seat width	min 43	48.22 - 50.32	50.10 - 55.66
Backrest height (Upper body Support)	min 65	53.92 - 67.66	52.29 - 65.58
Seatback width	20 - 32	17.57 - 27.37	19.25 - 28.70
Armrest height	min 28	23.78 - 31.61	21.33 - 29.20
Armrest length	min 47	41.47 - 62.06	40.20 - 64.52
Distance between armrests	43 - 46	39.70 - 53.48	38.47 - 55.03
Headrest height	NA	21.41 - 23.41	22 - 26
Headrest width	NA	13.66 - 16.08	13 - 16.24

*Dimensions are in cm

4. Discussion and Conclusion

It can be observed from Table 2 that several mismatches exist between pilot's seat design dimensions standards and the common North American and European anthropometric databases. To remedy this situation, several highlights can be made to be considered for determining new dimensions for current seats as follows:

- The maximum seat height in CAESAR dimensions is less than the dimension in AS290B and the minimum dimension in ANSUR is more than the standards. AS290B and the minimum dimension in ANSUR is more than the standards
- There is a mismatch between maximum dimension of seat depth in AS290B and CAESAR. Maximum dimensions based on ANSUR is 54.60 and 9.60 cm and is different with aerospace standards. In the standards, the maximum seat depth is less than the anthropometric found from both USA and Europe populations
- Backrest height obtained from ANSUR and CAESAR are based on shoulder height, which has a mismatch between the minimum and maximum height in AS290B standard
- Backrest width according to CAESAR and ANSUR data for shoulder width is different with AS290 standard. The minimum in AS290 standard is higher than 5th shoulder width while the maximum dimension is also less than the 95th percentile male in the CAESAR and ANSUR.

In conclusion, this study has established the notion that the design guidelines used for pilot's seat may not be adequate to cater for varied anthropometric characteristics of today's pilots. The mismatch should be taken into account to improve the design quality of pilot's seat in the future.

References

[1] Cîmpian, I. "Ergonomic Design of Aircraft Cockpit", Journal of Industrial Design and Engineering Graphics, 7(1), 2012
[2] Lusted, M., Healey, S. and Mandryk, J. A. "Evaluation of the Seating of Qantas Flight Deck Crew", Applied Ergonomics, 25(5), pp. 275-282, 1994
[3] Goossens, R. H. M., Snijders, C. J. and Fransen, T. "Biomechanical Analysis of the Dimensions of Pilot Seats in Civil Aircraft", Applied Ergonomics, 31(1), pp. 9-14, 2000

[4] Zhang, H., Zhuang, D. and Wu, F. "The Study on Pleasure and Ergonomics of Cockpit Interface Design", IEEE 10th International Conference on Computer-Aided Industrial Design & Conceptual Design, November 2009
[5] Drury, C. G. and Coury, B. G. "A Methodology for Chair Evaluation", Applied Ergonomics, 13(3), pp. 195-202, 1982
[6] Kennedy, K. W. "International Anthropometric Variability and its Effects on Aircraft Cockpit Design, Air Force Research Lab Wright-Patterson AFB OH., No. AMRL-TR-72-45, 1976
[7] Strickland, D., Pioro, B. and Ntuen, C. "The Impact of Cockpit Instruments on Pilot Exhaustion", Computers and Industrial Engineering, 31(1), pp. 483-486, 1996
[8] Salvendy, G. "Handbook of Industrial Engineering: Technology and Operations Management", John Wiley & Sons., 2001
[9] Choi, H. J., Zehner, G. F., Hudson, J. A. and Fleming, S. M. "Trends in Anthropometric Measures in US Air Force Aircrew Survey Data", Proceedings of the Human Factors and Ergonomics Society Annual Meeting, vol. 53, no. 10, pp.620-624, October 2009
[10] Marklin, R. W., Saginus, K. A., Seeley, P., and Freier, S. H. "Comparison of Anthropometry of US Electric Utility Field-workers with North American General Populations", Human Factors: The Journal of the Human Factors and Ergonomics Society, 52(6), pp. 643-662, 2010
[11] Aminian, N. O. and Romli, F. I. “Mismatch between Anthropometric Body Dimensions and Classroom Furniture in Malaysian University”, Proceedings of the 2012 Canadian Engineering Education Association Conference, Winnipeg, Canada, June 2012
[12] Karwowski, W. "Ergonomics and Human Factors: The Paradigms for Science, Engineering, Design, Technology and Management of Human-compatible Systems", Ergonomics, 48(5), pp. 436-463, 2005
[13] Reed, M. P., Schneider, L. W. and Ricci, L. L. "Survey of Auto Seat Design Recommendations for Improved Comfort", No. UMTRI-94-6, 1994
[14] Ismaila, S. O., Musa, A. I., Adejuyigbe, S. B. and Akinyemi, O. D. "Anthropometric Design of Furniture for Use in Tertiary Institutions in Abeokuta, South-Western Nigeria", Engineering Review, 33(3), pp. 179-192, 2013
[15] Bridger, R. "Introduction to Ergonomics", CRC Press, 2008
[16] "DEA 3250/6510 CLASS NOTES: Sitting and Chair Design", Cornell University Ergonomics Web, http://ergo.human.cornell.edu
[17] Karwowski, W. and Marras, W. S. (Eds.) "Occupational Ergonomics: Design and Management of Work Systems", CRC Press, 2003
[18] Stanton, N. A., Salmon, P., Jenkins, D. and Walker, G. "Human Factors in the Design and Evaluation of Central Control Room Operations", CRC Press, 2009
[19] Goossens, R. "Biomechanics of Body Support: A Study of Load Distribution, Shear, Decubitus Risk and Form of the Spine", Erasmus MC: University Medical Center Rotterdam, 1994
[20] Moroney, W. F. "Automated Anthropometric Measuring Devices for Use in Mass-Screening", Crew System Ergonomics Information Analysis Center Wright-Patterson AFB OH, No. CSERIAC-RA-98-002, 1998
[21] Harris, D. (Ed.) "Human Factors for Civil Flight Deck Design", Gower Publishing Ltd., 2004
[22] Gordon, C. C., Walker, R. A., Tebbetts, I., McConville, J. T., Bradtmiller, B., Clauser, C. E. and Churchill, T. "1988 Anthropometric Survey of US Army Personnel: Methods and Summary Statistics", 1989
[23] Harrison, C. R., and Robinette, K. M. "CAESAR: Summary Statistics for the Adult Population (Ages 18-65) of the United States of America", Air Force Research Lab Wright-Patterson AFB OH: Human Effectiveness Directorate, 2002

Women's Environmental Literacy As Social Capital In Environmental Management For Environmental Security of Urban Area

Donna Asteria[1], Herdis Herdiansyah[2], I Wayan Agus Apriana[2]

[1]Communication Department, Faculty of Social and Political Science, Universitas Indonesia, Depok, Indonesia
[2] Postgraduate Program, Universitas Indonesia, Salemba, Indonesia

Emai: donnaasteria@gmail.com, donna@ui.ac.id

Abstract: This study is about experience of women's role in environmental management to raise environmental security and form of women's emancipation movement. Environmental concerns conducted by residents of urban women who become environmental activists based on environmental literacy. Because of that, women's experience in interacting with both physic and social environment have differences in managing the environment including managing household waste by applying the principles of the 3Rs (reduce, reuse, recycle) and their persuasive efforts on their communities. This is the key to achieving sustainable development by anticipating environmental problem and preserving the environment. This study is conducted qualitative research method and its type is descriptive-explanative. The result of this study is environmental literacy of women activist on pro-environment action in their community that has achieved spiritual environmental literacy. Environmental literacy may differ due to internal and external condition of each individual. Pro-environment activities conducted as a form of responsibility of environmental concern such as eco-management, educational, and economic action, by persuading residents to proactively and consistently continue to do environmental management and develop a sense of community in shaping the networks of environmental concern in local context for global effect.

Keywords: environmental literacy, women activist, environmental management, environmental security

1. Introduction

In the context of environmental awareness, women have an important role to achieve environmental justice. Environmental security should be achieved globally through environmental management activities locally and regionally. Efforts to prevent women from becoming victims of environmental issues require collaborative action through the ability develop women's networking. The women activists also contributed greatly as a force for peace movements in conflict areas around the world [1-3]..

Environmental contamination that occurs in urban area at the present day is very concern. Environmental problem cause air pollution, clean water issues, waste issues, and environmental sanitation. To overcome the environmental crisis, it requires action from the whole society to act environment friendly. Environment friendly can be performed by planting and maintaining environmental hygiene that can reduce pollution or waste that can harm human's health.

Human is an actor who decides how to act and change the situation but the choice to act is influenced by social and ecological condition of its environment. Access and control over natural resources must also be owned by women as part of human rights. Therefore, the empowerment of women will be enhance the ability of women to participate in environmental management. Behavioral and responsibility in environmental management will be achieved by increasing individual environmental literacy. Individual environmental literacy will underlie and strengthen individual motivation and environmental concern action in preserving the environment. Individual who has ability in environmental literacy will have more sensitivity to the environment. This individual will be able to identify environmental problem in their community and find its solutions. Additionally, this individual is able to actively participate in maintaining and preserving the environment.

The ability of society to do environmental concern is a solution in dealing with environmental issues that will improve the quality of people's life and environment. The study was conducted in DKI Jakarta, Indonesia, where the environment pollution happening in the areas of Jabodetabekpunjur (Jakarta and its satellite cities/suburbs) nowadays is very worrying. To deal with the environment crisis, it is necessary to have a concrete act from all the community to have an environmentally–friendly behaviour, by doing greening and maintaining the cleanliness of environment to reduce the threat of pollution or waste

Behavioral change in environmental management often requires block leaders. Block leaders is a name for people who volunteer themselves (work without getting a wage or payment) to distribute information of environmental conservation [4]. Block leaders served as a major motivator for people, by directly delivering information and inviting people who become people mover to participate in environmental management in their community. In Jakarta, independent environmental management has been started to do by mobilization that is performed by people's motivation. At several locations in Jakarta, people's mover is conducted by a group of woman or housewife who later on persuade people to participate actively in managing their environment. The action that carried out by woman is a social capital to achieve sustainable environmental management. Individual environmental literacy will be different depending on the ability to understand and intensity of individual interaction with the artificial environment in the city. The condition of artificial environment in Jakarta is not just a matter of open land limitations in high populated residence areas but is also a pollution problems that more increasing. The environmental literacy and the ability of women in their role to manage environment, especially in the household waste management independently as local knowledge is need to elaborated as part of the individual behaviour ecology. As a role model in environmental management actions, it needs to be noted that environmental literacy on women activist is caused by experience of their activity that has contribution on the preservation of Jakarta area.

This study is to give a picture of human capital mapping socially existing in the society in environmental management based on environmental literacy of women. For the environment policy and its relation to the gender aspect, this research contributes to give a picture of the household waste management which can be replicated in other communities through granting access and woman empowerment where the environmental literacy of women can support the ecological continuity and human resources in the urban ecosystem.

2. Theoretical review

In this study to explore the power of women with a unique interaction with the natural environment, used the concept of environmental literacy. Environmental literacy has a wider understanding than ecological literacy. Environmental literacy is not only related to ecological understanding but also understanding between human interaction with physical/nature and social environment. The emphasizing in environmental literacy is based on ecological principle including inter-relation between nature environment and social system. Ecological knowledge is not only to discover environmental issue. Environmental literacy consists of several integrated constituent elements, namely: awareness, knowledge, attitudes, skills, and behaviors [5-7]. In this study, the element skill differentiates from attitude element because of personal skill. Related to the integration of forming elements of environmental literacy, the individual ability can be divided into several stage/level as follow:

- Nominal environmental literacy as a basic ability in realizing and understanding environmental issue and problem that are communicated in the form of environmental information.
- Functional environmental literacy as a concept knowledge that is related to ecology and environmental issue. Individual is also able to think critically and deliver environmental message to other parties both oral and written.
- Operational environmental literacy as knowledge and skill of environmental concern in daily life. Individual is also able to collect and evaluate information in order to plan action and act environmental concern [8].

The concept of environmental literacy is positioned in the same level with the local knowledge. The concept of local knowledge is assumed as knowledge about environment that belongs to individual as a common citizen. The concept of local knowledge is knowledge about environment and territory, related to information about neighborhood. Environmental literacy has not been defined specifically based on the types of information and also specific context level [9].

In this study, environmental literacy redefinition is performed. Environmental literacy is individual ability in understanding physical environment and making concept of environmental problem and solution, also taking decision, environmental concern behavior, as a learning process result and exercise from the interaction with the physical and socio-culture environment. Environmental literacy context is related with gender, urban resident and the participation in environmental management. Knowledge element in individual environmental literacy is related to knowledge of environmental ethic. Alternative environmental ethic in this study is associated with eco-feminism. Eco-feminism becomes an alternative in the way of thinking of environmental ethic aspect that examines relation of masculine social institution and the destruction on physical environment[10]. Women's close relationship with nature is based on historical condition on the pressure of patriarchy institution and western culture domination in identifying women and nature related with women biological attribute that is a result of social construction[11-12].

Related with the individual's environmental literacy, determine several form of responsible behavior toward environment, namely:

- Eco-management (physical action) refers to the direct environmental action to physical/nature environment to take precautions and seek solution to environmental problem that occurs.
- Consumer/economic action refers to environmental action by using financial/aid support in taking precautions and finding solution to environmental problem.
- Persuasion refers to environmental action that is performed by individual or groups by approach other side to prevent environmental problem and seek its solution.
- Political action refers to politically environmental action (political process through activity of party, company, etc) to prevent environmental problem and seek its solution.
- Legal action refers to environmental action that support and implement law/policy that is valid and plan policy to prevent environmental problem and seek its solution.

Environmental management will support sustainable society achievement [13]. Environmental management has to be accompanied by society's awareness to keep learning and seek a change in a way of thinking and understanding ecosystem life value. Value that has to be possessed by society is related to love for future generation through partnership and cooperation. Human needs to develop their self realization to understand their self existence and seek a balance and safe condition; it is not only for themselves but also for the ecosystem. Woman is also able to motivate their surrounding society in local scale in their community to preserve environment in practical and concrete way. Woman's closeness with environment causes woman to become watcher to keep environment preservation in root grass level in local area [14]. Related to environmental literacy, as in Beijing

Declaration 1995, on 11th critical area of woman and environment, it needs knowledge digging and woman participation in environmental management and achieve sustainable healthy environment.

3.Methodology

This study is conducted with qualitative research methods, performed with field research to dig experience and insight of female mover on pro-environment action. Data collected with method that is used is indepth interview method to 6 (six) participant and used participant observation method to join participant's activities. Beside of that, documentary study from newspaper clipping and other study/research on environmental management based on community to support the data. Extracting data with indepth interviews conducted with probbing and always recorded with a recording tools, researchers also made a record of all the events on the field notes from observations and the additional information when at the study site.

Sampling techniques were conducted to obtain participants by snowball technique and the technique in choosing informant is purposive with the criterion female mover on pro-environment action based on assumption of independent and proactively in environmental management by citizen and woman's ability in forming a networking in their community which can develop healthier and more comfortable environment quality. This study have 6 participant which is using saturation principle, sampling not continues until the researcher recognises that no new data were forthcoming or information redundancy [15]. The location of this study is Jakarta; informant residence which is located in Lebak Bulus, Banjarsari-Cilandak, Rawajati-Pancoran South Jakarta area, Klender-East Jakarta, Rawasari-Central Jakarta, and Warakas-North Jakarta.

Analytical techniques used in this study is using thematic analysis to find pattern or theme on information of the transcript of interview. Analysis is performed through process of coding data and complex indicator (from participant's statement in verbatim transcript), then made relation between indicators, defining and naming theme, then generated descriptive themes, after that wrote integral elements into a form of conceptual scheme/diagram. This conceptual result is gained inductively from field data and deductively from previous theory and study. In this analysis process, no statistical analysis used [16-17]. The way to maintain the validity or feasibility studies done by triangulation through member checking and triangulation of data collection methods (interviews, observation and document study).

4. Results and discussions

Based on the analysis found a theme that shows indicators of literacy and environmental elements after confirmation of concepts and theories, it is known that the participants already have the ability to achieve spiritual environmental literacy. This capability exceeds the knowledge and skills because the participants have been able to make alternative solutions and choice of activities to address environmental problems. Commitment and responsibility underlying due to their ability to be part of the worship of God and do good for God's creation. This key words is known from the coding process in the verbatim transcript follows:

Participant number 1: "... this is our responsibility as a creature of God .... God entrusts the environment to be maintained ....."

Participant number6: ".. to preserve God's creation is worship"

The capability exceeds the element of operational environmental literacy described in the review of the previous theory section.

Women's environmental literacy led them to do pro-environment activities and persuade people to sustainably manage the environment. Their always want to share knowledge and experience to other citizen which is also under lied by spiritual values that they believe. Women's orientation of the pro-environment action are prioritizing their activity to maintain the viability of the family, that family became prosperous and healthy are the ones that underlie women's awareness and sensitivity to the environment. Women has a value of altruism, empathy, sympathy, and care to preserve the environment by making changes for the better and more respectful of the surrounding environment. In this study, the activity of environmental management that is performed by the participants is more focus on the management of the physical/nature environment (eco-management). It is known that there are differences in the environmental literacy of women activist; this is due to the internal and external conditions of woman. Urban area typology condition, which is in this study is the city neighborhoods and housing complex, also has relation with the establishment of environmental literacy.

In the context of Indonesian, environmental literacy of women also motivated by situational conditions which is external thing for women, the condition of the community in a social environment, perspectives on the environment, and implementation of various policies in both the PKK program (PKK stands for Pendidikan Kesejahteraan Keluarga-Family Welfare Education), as well as laws on the environment and waste management. While the personal circumstances of women is related to the individual's characters or value such as self-concept and self-efficacy (an individual's capability to assess and decide to perform certain actions), the value of religiosity, and orientation on family life. External and internal conditions as the description above, cause individual environmental literacy will be different. However, sense of caring, compassion, empathy, and sense of brotherhood to share each other, owned by each female informant instinctively. Various changes in both nature and artificial environmental that caused by policy in citizen housing arrangement that is not integrated in dealing with population density becomes a shaper of awareness for the informant to prevent environment degradation.

Compliance of the condition of the physical/nature environment with personal necessity will form the different experience and knowledge, and also different sense of place in forming environmental literacy. This difference can be optimized by personal empowerment with a local behavioral approach. This is because condition/situational factors that may cause environmental literacy do not achieve behavior of environmental concern. Increasing environment knowledge will not directly cause the behavior to be concerned about the environment because there is access, control, benefits, and will strengthen the participation of environment responsibility. The basic thing in environmental literacy is the skill in reading and writing so that it is easy in receiving information, particularly environment information. Information and its source access are important things to be considered in forming the individual environmental literacy.

Meanwhile, the empowerment effort that is performed by women activist who made the social cohesion that strengthen social security in the community of the citizen socially will be the social capital to achieve ecological sustainability. Mitchell explains there are three conditions that make social cohesion; it is the individual's commitment to follow the value that is contained in the society to change environmental condition, the desire to share and feel to become a part of the community. Participation of female citizens as actor of pro environment action is an important social capital for sustainable community[18]. The ability of woman is social capital where individual resource is able to mobilize others through relationships with others (community citizen) through the establishment of interpersonal networks in the context of community empowerment will be more success with an approach that adjusted with the type of residence (related to the condition of the physical/nature and social environment).

5. Conclusion

The fundamental thing from environmental literacy element is woman's knowledge, attitudes, and behavior of environment concern. Their focus on environmental problems and issues that occur is more specialize in the neighborhood where they live. Women who have spiritual environmental literacy has the ability not only the skills but also the commitment and managerial skills to solve environmental problems. Underlying condition of spiritual environmental literacy in the context of the findings of this study are the traditional values of the herbs and how to grow crops, religion, profession, and the condition of living environment including the type of settlement that is densely populated. Although the findings of this study in a local context, however, has indicated that women have the experience and the power to form a network to improve the environmental issues to achieve environmental security and sustainable development globally.

6. References

[1] Tindall D B, Davies S and Mauboules C 2003 Activism and conservation behaviour in an environmental movement: The contradictory effect of gender *Society and Natural Resources* **16** pp 909-932

[2] Zelezny L C, Chua P and Aldrich C 2002 New ways of thinking about environmentalism: Elaborating on gender differences in environmentalism *Journal of Social Issues* **56** 3 pp 443–457

[3] True J 2001 Feminism, in Scott Burchill, et al *Theories of International Relations* (Palgrave) pp. 231-276

[4] Utami, B D, Indrasti N S and Dharmawan A H 2008 Pengelolaan Sampah Rumah Tangga Berbasis Komunitas *Sodality* **2** 1 pp 49-68

[5] Loubser C P, Swanepoel C H and Chacko CPC 2001 Concept formulation for Measuring environmental literacy *South African Journal of Education* **21** pp 317-323

[6] Orr, D 1992 *Ecological Literacy: Education and The Transition to a Postmodern World.* SUNY Series in Constructive Postmodern Thought. State University of New York Press. (Albany. New York)

[7] Sin-Jang Hsu and Roth, R 1996 An Assessment of environmental knowledge and attitudes held by community leaders in the Hualien area of Taiwan *The Journal of Environmental Education* **28** pp 24-31

[8] Disinger J F and Roth C E 1992 *Environmental Literacy* OH: Education Resources Information Center/Center for Science, Mathematics, and Environmental Education (Columbus)

[9] Hares M, Eskonheimo A, Myllyntaus T and Luukkanen O 2006 Environmental literacy in interpreting endangered sustainability: Case studies from Thailand and Sudan *Geoforum* **37** pp 128-144

[10] Keraf S 2006 *Etika Lingkungan (Environmental Ethics)* (Jakarta:Kompas Publisher)

[11] Rocheleau D, Thomas-Slayter B and Wangari E (Ed) 1996 *Feminist Political Ecology: Global Issues and Local Experiences* (London: Routledge)

[12] Wiesner-Hanks M E 2011 Crossing borders in transnational gender history *Journal of Global History* **6** 3 pp 357-379

[13] Sugandhy A and Hakim R 2009 *Prinsip Dasar Kebijakan Pembangunan Berkelanjutan Berwawasan Lingkungan* (Basic Principles of the Policy of Environmentally Sustainable Development) (Jakarta: Bumi Aksara Publisher)

[14] APPB 2008 *Sinergi Perempuan Dalam Pembangunan Berkelanjutan* (Woman Synergy in Sustainable Development) (Jakarta: Djambatan Publisher)

[15] Curtis S, Gesler W, Smith G and Washburn S 2000 Approaches to sampling and case selection in qualitative research: Examples in the geography of health *Social Science and Medicine* **50** pp 1001-1014

[16] Braun V and Clarke V 2006 Using thematic analysis in psychology *Qualitative Research in Psychology* **3** pp 77-101

[17] Attride-Stirling J 2001 Thematic networks: an analytic tool for qualitative research *Qualitative Research* **1** 3 pp 385-405

[18] Miller GT Jr 2004 *Living in the Environment: Principles, Connections, and Solutions* (Canada: Thomson Learning)

Mitigation Policy Scenario of Space Debris Threat Related with National Security

Herdis Herdiansyah[1], Evy Frimawaty[2], Ahmad Munir[2]
[1] Graduate Program, University of Indonesia, Salemba, Indonesia
[2] Environmental Science, Graduate Program, University of Indonesia, Salemba, Indonesia
E-mail: herdis@ui.ac.id, hard_disk82@yahoo.de

Abstract: The development of air space recently entered a new phase, when the space issues correlated with the future of a country. In past time, the space authorization was related with advancing technology by many space mission and various satellite launchings, or it could be said that who ruled technology will rule the space. Therefore, the numerous satellites in the space could be a threat for the countries which are mainly located in the path of the satellite, especially in the equatorial region including Indonesia. This study aims to create a policy scenario in mitigating the threat of space debris. The results showed that although space debris was not threatened national security for now, but the potential and its impact on the future potentially harmful. The threats of orbit circulation for some experts considered as a threat for national security, because its danger potential which caused by space debris could significantly damage the affected areas. However, until now Indonesia has no comprehensive mitigation strategy for space matters although it has been ratified by the United Nations Convention.
Keywords: satellite, space debris, mitigation, policy scenario, national security.

1. Introduction

The issue of space development has entered a new stage where it correlated with the future of a country. Developed countries are competed to explore the space for each interest. Developed countries advanced far above Indonesia such as landing mission (United States, Russia, and China), space exploration for science, defence and security (America, China, India, Europe, Japan) and the utilization of space energy (Japan). However, despite the potential and mysteries which still under covered, the space debris threat potential is also concerned recently [1].

The increasing number of space objects such as satellites that circulate around its orbit either still active or have become space debris caused the high possibility of crash between them. It is different with other space objects such as meteor and etc, in addition this study is only focused on object that made by human beings and in this case is satellite. Space debris is a type of space object which has expired and space object in the form of fragment / flake / piece / dust as result of collisions between space objects [2]

This potential threat is going bigger when flake / piece / fragments fell to the earth's surface. It caused damage to plants, animals, soil, water, air, and humans from radiation of radioactive materials inside the satellite. Therefore, the threat of space debris that circulates around the orbit regarded as definite potential threat to a country [3]. In Indonesia, according to Herdiansyah [4] research findings showed that the Indonesian public perception assumes space debris as one definite threat in the future.

In 2010, there were nearly 1100 spacecrafts and rockets near GSO where the operated one was only about a third of them. According to data from the European Space Agency (ESA), during the world satellite era, there were total of 6.000 satellites launched into space and 1.000 of them are still operated till now and the rest them become debris[5].

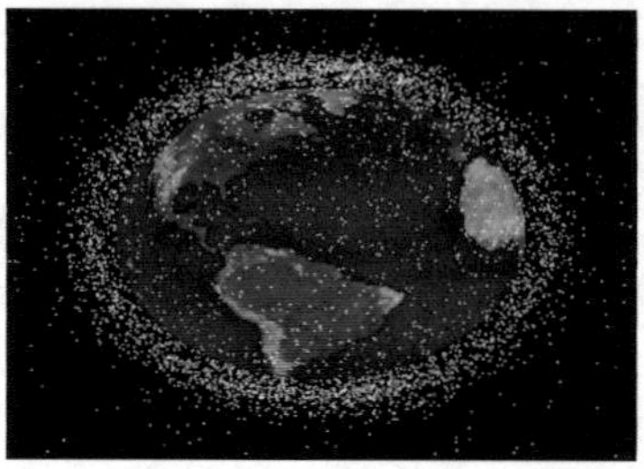

Picture 1.Space debris on the Earth's orbit

(http://www.nasa.gov/mission_pages/station/news/orbital_debris.html#.UrDBgScQQ-B)

The geographical of Indonesia has dangerous potential as the opportunity of space object falling is very high. Until now, the space objects were felt down in several times at some regions of Indonesia, even not caused significant losses and human life. Based on the data, there are three falling space objects and could identified as part of COSMOS-3M rocket motor which felt on March 26th 1981 in Gorontalo. The others are part of Soyuz A-2 rocket motors which felt on April 16th 1988 in Lampung and the broken of CZ-3 PRC rocket which felt on October 13th 2003 in Bengkulu [6]

2. Theoretical Review

Space Debris can be defined as all human-made objects that orbiting on the space of earth and had no benefit for human. Space debris is moved in orbit with certain speed before finally entering (re-entry) the Earth's atmosphere [7] [8]. Space debris are in form of falling objects such as satellite, a former rocket, or satellite either rocket fragments. Commonly, space debris is threatened for its collision effect with atmosphere and / or with surface of the Earth that consist harmful material such as nuclear radiation [9].

Moreover, in 2011, the significant increasing of solar activity affected in the increasing of space debris number that felt to the Earth. Theoretically, the correlation between solar activity and the amount of falling debris could be easily seen if we only considered with the former satellite and rocket (ignoring the small-sized flakes) [10]

Because of many space debris felt as increasing of solar activity since 2009, the amount of space debris on the orbit was decrease in year 2011. This trend will continue to 2013 if there is no explosion of satellites or used rockets in large numbers. It will increase after 2013 and become threat for the countries in equatorial region.

The issue of space debris becomes part of the national security threat since the development of international relation where strengthened in one side but weaken in the other side. At first, the national security which related to the relationship between states was considered too focused on studies of high political, intelligence, and natural resources competition. It means that the national security in international relation dimension is concentrating on state, people, international system, and reciprocity contribution [11]

After the end of Cold War, national security was transformed on the issues that have no big impact directly. Hence in conceptually, national security in the former was known as militaristic approach and dominated with issues of defence, security dilemma, arm race, nuclear, weapons, etc. Recently, the national security threat changes the characteristic of threat to become more complicated. Therefore, the security problem becomes more comprehensive because involving other aspects such as economy, social-culture, environment, and even public health issue [12]

Security in simple traditional concept could be defined as a state security that could be threatened by another military power and should be defended by its own military strength [13] [14]. One of those points is the development of satellite orbit that threatens the majority of countries which are not the owner of those satellites. Therefore, the national security which related with satellite orbit is basically trying to discuss how a country faced the threat of space debris.

3. Methodology

This study design followed the working of mixed method approach that proposed by Creswell [15], where the research will be accomplished by quantitative and qualitative approaches sequentially.

Although used mixed method, this research also used various methods that adapted with the purpose of research in its implementation. Quantitative approach is used to test the existing construction, while qualitative approach is used to explain and describe the phenomenon of the theme.

The type of this research is descriptive analytic in form of research that provide picture or description of phenomenon without any treatment to the object which is studied analytically according to the factual findings on the field. However, intervention is done by modelling and specific explanation of data, hence it will create an applicative model that could explain the theme.

The populations on this study are the stakeholders in the issue of satellite orbit circulation as well as space debris, such as government, satellite private user, NGOs, academia, and the satellite user community. The criteria of informant are determined by factors such as its expertise, involving, and having strong track record. The qualitative informant consists of regulators in this case of Communication and Information Ministry and National Institute of Aeronautics and Space (LAPAN). Private sector consists of several telecommunication companies that have their own satellite. The working steps are starting by gathering relevant data, formulating the problems, carrying out preliminary studies and theoretical framework, collecting data, carrying out intensive interview, and arranging scenario. The next are interpreting and analyzing data, formulating model, concluding and arranging final model or alternative policy related to the threat of satellite orbit.

The methodology of this study is started by collecting secondary data related to the study and goal of the research. Then researcher will formulate the problems and correlated it with the objectives of study. In supporting the analysis, the researcher will conduct former study and literature review. The next step is collecting data by two methods. Those are intensive interview and focus group discussion (FGD). Then the researcher will make interpretation and analyze the data. After that, the model could be formulated based on data analysis. Finally, researcher will make conclusion of the study and final model of alternative policy based on the analysis of this study.

On the other hand, the informant of this study will describe on the table below.

No	Institutions	Informant criteria
1	LAPAN director of space affairs	LAPAN leader whose has duty in space objects affairs
2	LAPAN researcher of space affairs	Government researcher in space affairs and focused on science and space objects
3	Academician in satellite study	Academician in satellite objects and master in operating and building of satellite
4	Academician in system thinking expert	Academician who focused on dynamic system modelling
5	Academician in national security matters	Academician in national defence and security and experienced in national security issue
6	Regulator of satellite arrangement	Representation of government who has authority on satellite operation in Indonesian space orbit territory
7	Kominfo researcher about frequency	Representation of government researcher

This study is descriptive analytic study in form of research to describe about object without any analytic execution of the object accordance with factual field findings. Qualitative approach method is used to explain and describe the theme phenomenon. Data resources of this study are in form of primary data and secondary data. Primary data was obtained by intensive interview as well as focus group discussion (FGD). Through compilation of those data resource, the result expectation is research result based on real phenomenon and oriented in applicative research. Hence the formulation of policy scenario is based on empirical and authentically facts [16].

4. Results and discussions

In the context of national security, the principle of policy scenarios in facing threat can be carried out by : 1) diplomacy for finding allies and isolating threat, 2) using economic power to force cooperation, 3) maintaining the effective armed forces, 4) performing civil defence and emergency readiness, 5) ensuring rapid recovery and increasing critical infrastructure, 6) using intelligence services to detect, defeat, or avoid threat and espionage, as well as protect confidential information [17]. Therefore, in facing the circulation orbit threat, the national security system is not only based on military force but also involving and integrating various non-military strength and institutional cooperation as well as responsive foreign policy. Indonesia which geographically located in the equatorial region along 1/8 of the earth, which is from 91 ° E to 141 ° E and width of 6 ° N to 11 ° S, is a highly strategic region for spacious activities. Refers to economic aspect, the launch of spacecraft would be more beneficial if carried out from the territory of Indonesia which is located just below the equator. This comparative advantage sometimes become target from developed or even developing countries which actively conduct spacious activity to conduct various spacious researches and launch spacecraft from the territory of Indonesia (for instance in Biak). However, its negative impact is relatively not to be thought [18].

The problem of space debris put national security issue as separate study. The context is the development of interaction between actors in management of satellite orbit such as damage potential and the compensation. In the development of national security issues, the classification of "security" definition is not only about militaristic and territorial invasion (traditional threats) but also getting complex due to satellite circulation orbit that threaten certain countries. In the definition of national security studies, impact and potential threat of non-traditional relationships in the region are caused by several problems such as vulnerability due to natural damage from space debris as well as linked with the wider social political economy interests.

The countries that have same possibility as location of the falling space debris could make this detriment issue as bargaining power with the owner of satellite and result in the higher of interaction of each countries as well as regional interaction. The expansion of national security threat is happened due to the danger and impact of the falling space debris is getting wider when it contains hazardous materials such as satellite Cosmos 954 with weight of 4.5 tons and contains radioactive substances that fall in Canadian sea on January 24th 1978. Its radioactive substance spread along 600 km from Great Slave Lake to Baker Lake [19].

The United Nation thru UNCOPOS (United Nations Office for Outer Space Affairs) has formulated mitigation guidelines for space debris which consist of [20]:

1. Restricting the release of space debris in normal operation.
2. Minimizing the pieces on space object operation. Therefore, since the beginning, it should be prepared a planned anticipation, such as: mechanism of energy abolition at space system (passivation) and disposal system,
3. Restricting the accidental collision in orbit for increasing of space debris volume and mass,
4. Avoiding intentional destruction of space object or other dangerous activities,
5. Minimizing the potential fragments that caused by the rest of energy after the end of space object operation,
6. Restricting long-term disruption in LEO and GEO orbits that caused by the end space object operation.

The scenario of threat mitigation policy is started from 3 approaches. First is hazard assessment approach. In this phase, the threat mitigation is needed to identify the potential damage and the level of threat that might be happened. In dealing with the threat of satellite orbit circulation, hazard assessment phase requires knowledge and high technology in order to map the worst possibility. The outcome of this phase is the data of regions that may affected by the circulation of satellite orbit [21].

The second phase is warning. This phase is happened when the threat of space debris has to be real. Then the process of mitigation operates in form of warnings based on available data. Mitigation process in this phase is done in quick and carefully ways to avoid victim. The third phase of mitigation process is called preparedness phase. In this phase, the knowledge of impacted area due to space debris and the knowledge of warning systems of evacuation and consolidation when the situation has already safe are highly required. Those three mitigation process are basically preventing and minimizing damage and victims.

LAPAN (National Institute of Aeronautics and Space) in 2009 has developed monitoring software of falling space objects which later became part of the information system and dissemination of falling space objects that developed in 2010. The device named "Track-it" recently is able to automatically display the trajectory objects that existed on USSPACECOM catalogue when it passes the territory of Indonesia with height of less than 200 km in one hour before and one hour after.

"Track-it" is a program to facilitate the monitoring of artificial objects that orbiting the Earth and potentially fall if its height is quite low. Generally, the object that less than 200 km height will fall in less than 2 days. In an altitude of 122 km is commonly considered as critical height of a falling space object. The objects that less than its critical height are usually expressed falls (atmospheric re-entry condition) even the crash site on the Earth's surface could not be known. Therefore the public is able to know the identity of object that had just fallen or will fall in Indonesia territory [22]. However until now, LAPAN (National Institute of Aeronautics and Space) has no mitigation guides and strategies about how to deal with the falling of space debris in Indonesia.

According to LAPAN informant, the numerous satellites in equator territory caused Indonesia in dangerous area refers to space debris. In addition, many developed countries ignored the compensation of resulted space debris. Therefore, the country that passing by its space debris such as Indonesia should has anticipation planning and steps. McCall et all [23] said that this situation called space situational awareness (SSA). It is important for any country which has concern military and economic well-being. SSA is the enabling of a description of the location and operation of space assets. SSA in US identifies the capabilities needed to keep US assets and destroy those kinds of the enemy. Kind of mission was conducted to destruct the space awareness but not violate the treaties or grounds for retaliation.

According to national security informant, the threat of space debris for national security is not just in real and definite threat. The impacts were not significantly happened now and therefore some of the countries ignored about this issue. However, Rusia and China have made special force for space as its great potential empowerment and the danger of space debris that increase during the increasing of new satellites and the old satellites which still exist in space till down to earth. McCall etc [24] said that threats from space debris objects or uncontrolled satellites are passive threat. The most important factor of SSA (space situational awareness) is always location. Orbital parameters can calculate its location measurement, therefore it can search the objects that can be dangerous to one country's space assets. It also generates warning at proper times to encourage defensive actions. Other passive threats are from high-energy particles and photons which caused by natural impacts such as solar storms or impacts of nuclear explosions in the atmosphere. Therefore, the detection and calculation in dealing with those space assets should be determined by characteristics and possible dangers of its threats.

5. Conclusion

It can be concluded that space debris which potentially fall down from the orbit to Indonesian territory could be a threat for Indonesian citizens or even become disaster that caused human victims. In wider perspective, as its potential of threat and its level of damage, space debris becomes part of national security threat and forced us to arrange policy scenario in facing it. One form of mitigation against the threat of space debris is how the mitigation system could avoid the fall of satellite or space debris and could avoid damage and victims. Although in the space matters context, the space threat mitigation is obligation of all countries especially the owner of satellite, but the countries which are located in equator and threatened area have higher interest rather than the countries beyond the equator. This mitigation process is important and became one of the objectives of this study. The scenario of threat mitigation policy is started from 3 approaches that the hazard assessment, warning action based, and preparedness phase. Those three mitigation process are basically aiming to minimize damage and victims caused by space debris fall down. Therefore, the policy scenario should be able to assess the danger (hazard assessment), to know when the warning is activated (warning), and to prepare the threat anticipation (phase of preparedness). Unfortunately until now, the developed countries as the largest satellite contributors in space are not much concerning with compensation of space debris that fall down into other countries, even though the policy level of United Nations has made guidance of space debris management.

ACKNOWLEDGMENTS

This paper is national collaboration research and was supported with Puslitbang SDPPI Kominfo RI

6. References

[1] Bedingfield K.L Leach, R.D and Alexander, M.B. 1996 *Spacecraft System Failuresand Anomalies Attributed to the Natural Space Environment* (USA: NASA Reference Publication)

[2] Button M. 2013 Cleaning Up Space: The Madrid Protocol to the Antarctic Treaty as a Model for Regulating Orbital Debris (USA: Wm. & Mary Envtl)
[3] Hildert, Steven A and Allison Arnlod 2014. Threats to U.S. National Security Interests in Space: Orbital Debris Mitigation and Removal (USA: Congressional Research Service)
[4] Herdiansyah Herdis. 2010. Survey Persepsi Publik Tentang Keamanan Nasional (Jakarta: CSDS-Ditjen Pothan unpublish)
[5] Yuniarti Diah, 2013 Analisi SWOT Sampah Antariksa Indonesia *Buletin Pos dan Telekomunikasi***11** 1 pp 13-28
[6] Susanti Dini 2014 Analisis Peran Kelembagaan LAPAN Dalam Mitigasi Ancaman Keamanan Nasional di Bidang Keantariksaan Terkait Dampaknya Perubahan Iklim. (Jakarta: Thesis Program Pascasarjana Universitas Indonesia)
[7] Neflia E 2010. Upgrading Sistem Diseminasi Informasi Benda Jatuh Antariksa Dari Temporer Menjadi Mingguan (Jakarta: LAPAN)
[8] Yuniarti Diah, 2013 Analisi SWOT Sampah Antariksa Indonesia *Buletin Pos dan Telekomunikasi***11** 1 pp 13-28
[9] Imburgia Joseph S 2011 Space Debris and Its Threat to National Security: A Proposal for a Binding International Agreement to Clean Up the Junk *Vanderbilt Journal of*Transnational Law **44** 589
[10] Neflia E 2010. Upgrading Sistem Diseminasi Informasi Benda Jatuh Antariksa Dari Temporer Menjadi Mingguan (Jakarta: LAPAN)Neflia E 2010. Upgrading Sistem Diseminasi Informasi Benda Jatuh Antariksa Dari Temporer Menjadi Mingguan (Jakarta: LAPAN)
[11] Imburgia Joseph S 2011 Space Debris and Its Threat to National Security: A Proposal for a Binding International Agreement to Clean Up the Junk *Vanderbilt Journal of Transnational Law***44** 589
[12] Susanti Dini 2014 Analisis Peran Kelembagaan LAPAN Dalam Mitigasi Ancaman Keamanan Nasional di Bidang Keantariksaan Terkait Dampaknya Perubahan Iklim. (Jakarta: Thesis Program Pascasarjana Universitas Indonesia)
[13] Herdiansyah, Herdiansyah Herdis. 2010. Survey Persepsi Publik Tentang Keamanan Nasional (Jakarta: CSDS-Ditjen Pothan unpublish)
[14] Susanti Dini 2014 Analisis Peran Kelembagaan LAPAN Dalam Mitigasi Ancaman Keamanan Nasional di Bidang Keantariksaan Terkait Dampaknya Perubahan Iklim. (Jakarta: Thesis Program Pascasarjana Universitas Indonesia)
[15] Creswell John W 1994 Research Design-Qualitative and Quantitative Approaches *(*NewDelhi: Routledge).
[16] Moleong, Lexy J. 2005. *Metodologi Penelitian Kualitatif Edisi Revisi,* Bandung: Remaja Rosdakarya.
[17] Susanti Dini 2014 Analisis Peran Kelembagaan LAPAN Dalam Mitigasi Ancaman Keamanan Nasional di Bidang Keantariksaan Terkait Dampaknya Perubahan Iklim. (Jakarta: Thesis Program Pascasarjana Universitas Indonesia)
[18] LAPAN 2011 Penelitian Sampah Antariksa di LAPAN: Bidang Matahari dan Antariksa (Jakarta: LAPAN)
[19] Neflia E 2010. Upgrading Sistem Diseminasi Informasi Benda Jatuh Antariksa Dari Temporer Menjadi Mingguan (Jakarta: LAPAN)
[20] United Nations: Office for Outer Space Affairs 2010 Space Debris Mitigation Guidelines of the Committee on the Peaceful Uses of Outer Space (UN: Vienna).
[21] Carter W Nick 1992. *Disaster Management : a Disaster Manager's Handbook* (Manila:Asia Develompment Bank)
[22] Susanti Dini 2014 Analisis Peran Kelembagaan LAPAN Dalam Mitigasi Ancaman Keamanan Nasional di Bidang Keantariksaan Terkait Dampaknya Perubahan Iklim. (Jakarta: Thesis Program Pascasarjana Universitas Indonesia)
[23] McCall, Gene H; Darrah, John H. 2014. Space Situational Awareness: Difficult, Expensive-and Necessary. *Air & Space Power Journal 28.6* (Nov/Dec 2014): 6-16.
[24] McCall, Gene H; Darrah, John H. 2014. Space Situational Awareness: Difficult, Expensive-and Necessary. *Air & Space Power Journal 28.6* . 6-16.

IOP Conf. Series: Earth and Environmental Science **30** (2016) 012016 doi:10.1088/1755-1315/30/1/012016

Littoral and Coastal Management in Supporting Maritime Security for Realizing Indonesia as World Maritime Axis

Agus Brotosusilo[1], I Wayan Agus Apriana[2], Afrizal Agung Satria[2], Trisasono Jokopitoyo[2]
[1] Faculty of Law, Universitas Indonesia, Depok, Indonesia
[2] Postgraduate Program, Universitas Indonesia, Salemba, Indonesia

Emai: broto.susilo@ui.ac.id, agus_brotosusilo@yahoo.com

Abstract: The Indonesian under President Joko Widodo has new goal to make Indonesia as the world maritime axis. This is supported by the geographic of Indonesia as the largest archipelagic country where the sea is two-thirds wide among the whole spacious. Indonesia is the world largest archipelagic state. More than two-third of its territory consist of seas. The ecosystem of littoral and coastal has correlative relationship with country development. There is no doubt of physically facts that Indonesian littoral and coastal with total wide of 5.8 million km^2 is rich with various natural resources. Therefore, the condition of Indonesia with its world second longest coastline has several comparative advantages. Not only the country has an abundant natural resources, but it also blessed by demographic bonus advantage. The population of Indonesian is the fifth largest in the world which approximately 220 million people and approximately 60 percent among them live at coastal areas. The people in coastal area relies their live from its surrounding natural resource. Hence, most of their life and daily activity is related with the presence of natural resources. The dealing of conflict potential and attention to maritime security are important to be studied as a reference in preparing and facing the government policies that will lead to the development of maritime.

Keywords: natural system, coastal management, maritime axis, littoral areas, security conflict.

1. Introduction

Indonesia is the largest archipelagic country. It consist of 17,508 islands and 81,000 km coastline as well as numerous potential littoral and coastal resources. [1]. The littoral and coastal resource consists of renewable resource such as fisheries, mangroves, and coral reefs. Meanwhile un-renewable resources are oil, gas, mineral, and environmental services [2]. For example in economy, the Indonesian fishery potential resources is estimated 6.7 million tons per year, where 4.4 million tons is from territorial and Indonesian waters, and 2.3 tons is from Exclusive Economic Zones (ZEE) waters (Department of Marine and Fisheries, 2002). Meanwhile the coastal areas are rich of mining and mineral as well as its potential to develop industrial activities, tourism, agriculture, housing, etc. The entire economic value of coastal and marine resource potential is up to 82 billion dollars per year [3].

Indonesia as the world maritime axis is supported by the wide of littoral and coastal areas. Coastal area is an area of land that is adjacent to the sea, with its limits of land includes flooded or not flooded areas and still influenced by ocean processes such as ebb and others. The Coastal area is dynamic and susceptible with environmental changes either by natural processes and human activities. The coastal as transition area between land and ocean ecosystem is an area on the surface of the earth's that mostly densely populated. Nearly 60% (as happened in Indonesia) of the world total population are living in the area around 60 km from the coastline. Two thirds cities in the world with more than 2.5 million populations are located in coastal areas. Even in Australia, 86% of its population is located in coastal areas [4]. According to geographical issues as archipelagic country, Indonesia has potential conflict with neighboring countries as well as maritime security issues that recently influenced national policy to develop maritime sector.

As consequence of Indonesian strategic position, the ocean of Indonesia is very important for the world community marine users. If Indonesia could take advantage of this challenge and opportunity, it will improve the nation's welfare. Based on history note, the ancestor of Indonesia was able to control the sea of *Nusantara* (Indonesia in the past), moreover they could voyage the wide ocean till the coast of Madagascar, South Africa. It was proved that the Indonesian ancestors had maritime character in cooperated with other countries in the world. In addition, the Indonesian ancestors had understood the meaning and the useful of sea as a means to secure its national interests, such as trade and communication. Therefore it can be concluded that the traditional use of sea was for transportation and the world trading is mostly passing the sea where its volume is going increase until now [5].

2. Theoretical review

The condition of littoral and coastal areas has impact of resource utilization by its communities which is very dynamic and highly dependent on availability of its surrounding resources [6]. Coastal community in generally have become a part of pluralistic society, however, it is still have togetherness spirit. It means that the structure of coastal communities is combination of urban and rural community characteristics. The coastal community structure is various, hence it forms system and cultural values that is culture acculturation of each components that formed the society structure.

Coastal environment is dynamic, unique, and vulnerable to environmental change [7]. Factors that influenced the coastal environment are including: population growth, climate change, the increasing of space and resources demands, and beach variety. The annually increasing of population growth that lives in coastal areas caused the increasing of human activity especially in utilization of natural resources and coastal ecosystem. Various kinds of human activities both in land and sea are stimulating its coastal environmental change including its coastline [8]. Therefore, as Lanza said about beach replenishment, careful management of resources will allow the reduction of management costs for maintenance, which has always been the critical point of artificial beach replenishment [9]. It also meant that careful management is crucial for environment matter in this case is for littoral and coastal environment.

Human activities such as mangrove clearing, sea sand mining, and coral reefs mining in several locations have contributed significantly erosion of the coast, because the losses of coastal protection from waves and hurricane, directly or indirectly, it reduces sediment availability on the beach if it compares with the losses of sediment because of natural impact. This is probably caused by continuously huge waves from the sea or perpendicular to the coast which finally influences the people's pond under the coastline of redundant beach. One of the most influential waves is the sharpness of the wave that usually occurs during high winds or storms which resulted in much coastal erosion.

Meanwhile in coastal management, the model favors the involvement of stakeholders, but does not use complicated assessment procedures for non-economic indicators or relative weights to combine economic, social, and environmental indicators. Instead, the integration between economic activities or properties and the environmental status and landscape is represented objectively and non-linearly by referring to initial and sustainability conditions, combined with budgetary and environmental constraints. The model successfully accounted for both human and environmental dynamics by depicting delayed effects, neighborhood externalities, and feedback effects [10].

The logical consequence of Indonesian water geographical condition is having political value and security strategy not only for Indonesia but also for other countries especially in Asia Pacific region [11]. The existence of maritime security challenge from the possibility of conflicts among states is becoming concerned issues. The conflict refers to the competition level among states in requiring natural resources and claiming of national and territorial boundaries.

Based on Bateman [12], Maritime security is a major concern in the Asia Pacific, yet is riddled with highly complex "wicked problems" that have thus far defied solution. These include different interpretations of the Law of the Sea underpinning regional maritime security, the lack of good order at sea, numerous conflicting claims to maritime jurisdiction, the implications of increased naval activity in the region and the lack of agreed maritime boundaries.

According Londynsky of Matson Navigation [13], in order to improve maritime security, two things must happen. First, The Department of Homeland Security must find a way to share actual threat information with the carrier community so that the true threat can be quantified. This will assist us in better allocating our resources. Second, there needs to be more coordination between security grant recipients so that projects can build on each other and be managed in such a way as to capture the best ideas. These ideas will then be able to build on each other toward a more complete security solution. Security is very important in this case, as well as detailed by Casey [14], who said that Singapore concerns with security on Mallaca Strait as transnational line of sea transportation. Therefore, Singapore has placed significant emphasis on the maritime threat environment. This is caused by potential of conflict that related with national strategic matters such as strategic interest in military and economy. Furthermore, this is caused by the great resource of its ocean such as crude oil and gas. The example is at around South China Sea where having gas potential for almost 18 trillion tons. Therefore this sea becomes source of conflict between neighboring countries that could endanger the maritime security of Indonesian [15].

According to Londynsky [16], the carrier communities understand that they have a unique responsibility to maintain the security of our global supply chain. There is no question that maritime transportation is a particularly attractive target to those who hope to do us harm. It is clear that an attack on the maritime sector could have an impact on economic viability of USA and at the same time potentially impact the ability to project power overseas. In addition, according to Gomes et all [17], number of strategies can be suggested for the improvement of conditions in the coastal area. One of them is the regulation of land use in order to minimize conflicts between the economic sector and protected areas.

The role of local government in optimizing the local coastal communities and stakeholders is very important to create together comprehension of maritime security. It also said by Helmick [18] that a "top-down" approach to the crafting of a long-term and sustainable research agenda in maritime security would be inappropriate and wasteful. Industry, organizations, state and local governments, the academic community, and other constituent groups represent enormous collective expertise and in many cases, are the entities that are ultimately responsible for the deployment of new technologies and the implementation of new procedures intended to enhance security.

3. Methodology

The study used qualitative research construction. According to Usman and Akbar (1996), the objective of qualitative research is not getting relationship between variables but just looking the meaning. On the other hand, based on Surakhmad (1985), the correct description of phenomenon in qualitative research is not limited just only in collecting and compiling data, but should include data analysis and interpretation of data's meaning.. The research steps and stages are done by observation and intensive interviews with informants in Jakarta, Manado, and Medan. The selection of study sites is based on the fact that the major ports and coastal area are numerous in these regions. Medan city in North Sumatera which controlled Belawan port city becomes monitoring and supervisory traffic postal in Malacca Strait. Malacca Strait is a very busy because the numerous of its maritime's traffic, hence that potential of problems and conflicts is very large along that area. The researcher hopes to get example of maritime problems and conflicts of sea transnational traffic in order to arrange the best cooperation model as the objective of this research. The other city is Manado in North Sulawesi which controlled Bitung port city. Its position as transit and service port to the ships that passing ALKI II (Indonesian chokepoints) makes Indonesia is important to have power of diplomacy and strong maritime cooperation.

4. Results and discussions

Coastal area is an area with the highest interaction of land and sea. This area has relationship with upland and the broken of it will have negative impact to the shoreline. For example, agriculture and forestry activities without concerning the conservation rules will cause erosion and flood as well as damage the ecosystem of river and continues to the coastal and coastline. Other example is people's activities which produce industrial and household waste that causing sedimentation and flow through the river to the coastal area. Brotosusilo [19] said that any change of landscape and its negative impacts such as pollution, erosion, and drastically changes of freshwater regime in land ecosystem will have impact to coastal ecosystems, including the coastline.

According to the result of observation and collecting information from informant in Manado, North Sulawesi, there are some findings. Manado is the city of North Sulawesi, meanwhile the port city of this area is Bitung which located about a hour from Manado city. Bitung is preparing for the big port city where its port will be prepared as the North gate of Indonesia. Therefore the atmosphere of maritime there is so signed. Based on the city situation, it has been arranged to become port city with well supporting of infrastructure. Shipping industry is running together with the progress of the fisherman who has small scope of getting income from the sea. Many people there are working in fishery canning industry. The local culture is also shown maritime attitude by its primary food which are not far from fish. Besides economic port that exist there, it also shoreline in some part of Bitung city. That place was used by the fisherman to place their boats and selecting their fish before send to bigger collector. Some beach is also becoming tourism destination for surrounding people and even the people from Manado. It seems that the atmosphere of Bitung Port City as maritime centre in northern Indonesia influence the people to act and develop their live with maritime comprehension.

The comparison between findings and previous research result could be explained by the table below.

Description	Previous research result	Finding
Ecological of surrounding coastal ecosystem	Shows worse condition as pollution, mal-treatment, and impact of global warming.	The condition is still getting worse but many programs conducted to make it slower and even many conservation programs aimed to avoid further breakdown.
Developed natural resources by the people	There are many obstacles to maximize the people's income on coastal area. Those are coming from outside (illegal fishing) and inside (minimum concern of fisherman living from the government).	The change of doctrine of the country results in initiative effort to develop maritime sector including develop natural resources for fisherman welfare and maritime stakeholders goodness.
Empowerment of groups that rely on natural resources	The empowerment was running well but in-maximum speed to obtain the specific nor great goals from maritime sector	The empowerment is forcing to be fast in order to reach high target of getting maximum advantages from maritime sector. It also supported by government maritime policy which stand behind the national interest.
Maritime perspective of people	Low consciousness because maritime thinking was covered by land orientation	Getting raise along with government campaign of being the world maritime axis
Maritime Infrastructure	Weak and unsupported the willing of becoming maritime country	On progress of revitalization and expansion of capacity instead capability.
Littoral and coastal development	Running slow as many problems that have no serious mitigation	Getting run faster and shows definite progress on the right track.
Coastal people living	In poverty and have no spirit of becoming prosperous by just only depends on coastal surrounding	Getting prosperous during the development steps of making coastal live quality better than before.

Government mission	Maritime is not priority, even though the state was getting significant income from maritime sector, but it was not maximum	Maritime as vision and goal, therefore the development of infrastructure and maritime consciousness is forcing faster than the others.
Achievement of mission recently	Giving great positive income to the state but not as maximum as we hope	Beginning to see by the decrease of illegal fishing meanwhile the increasing of coastal people and fisherman contributions are still forcing to the max.

The gap analysis of this research is describing based on the findings of research. Based on literature review and crossed with interview result, the former situation about littoral and coastal live in Indonesia is very pity in general. Perspective of mostly people at the former was mostly filled by land orientation, hence the sea potential was ignored to be maximized. It has been general knowledge that the fisherman and the people who live in coastal area were living in the low level of prosperity. Meanwhile our sea potential such as fish was stole by another country's fisherman who had modern and big ship to catch the fish. The local fisherman could not do anything because they defeated by big ship and the law enforcement of that violation was so weak. The infrastructure problem such as commercial port and other supporting tools were also became obstacle of maritime side development. The coordination and cooperation between related maritime stakeholders which have existed for a long time, practically brought problems that resulted in in-optimum work of developing maritime potential as well as maritime security. Abdulkadir [20] that the idea of preventing abuse of global market should never be seen as being incongruous with economic interests because security cannot be compromised for economic advantages. Therefore, formulation of maritime or ocean with laws is for the administration of maritime sector. For ensuring economic gains, it is not compromises with national security, which could be effected through environment insecurity

Since the new President Joko Widodo who promoted the mission of Indonesia as maritime axis power in the world, the situation was changed slowly but sure. Meanwhile the law enforcement of illegal fishing is showed bravely by bombing the ship that has been proved stealing fish in Indonesian water territory. Slow but sure, the direction of maritime development in Indonesia is getting clear. It shows by the observation of researcher in Bitung and Belawan port cities. The people who are living in coastal area begin to develop their sightseeing of maritime axis by supporting the government effort to enrich their fish, maritime production, and even maximized its maritime tourism potential to develop the local prosperity and burned out the spirit of maritime country. The mindset, attitude, and nation pattern should be based on maritime spatial awareness of where we are. Therefore maritime vision becomes a need for Indonesia [21].

5. Conclusion

Littoral and coastal aspect to support maritime security is basically used to avoid any conflict of maritime that related with space seizure. Based on geo-politic perspective, there is an interaction between the space and human beings. That interaction results in space consciousness which directly or indirectly related with security and human prosperity interests. Space awareness concept which originally started from individual and family level is developing to the wider aspect. In the context of modern state, spatial awareness concept is realized by the claim of sovereignty, which limited with boundary and laws as well as apparatus to ensure security and sovereignty.

Management of littoral and coastal in supporting maritime axis is basically in accordance with the vision of the present government who wants to transform and build maritime axis as a vision or goal of nation. In this context, the idea of the maritime axis is a call to Indonesia to return to national identity as archipelagic country, which will build Indonesia as united maritime power, prosperity, and dignity. Maritime axis is a doctrine that guides us to a sense of common purpose.

6. References

[1] Bengen DG 2001 Sinopsis Ekosistem dan Sumber daya Alam Pesisir dan Laut Bogor: Pusat Kajian Sumber daya Pesisir dan Lautan Institut Pertanian Bogor

[2] Dahuri R J Rais SP Ginting dan MJ Sitepu 2001 Pengelolaan Sumber Daya Wilayah Pesisirdan Lautan Secara Terpadu Jakarta: PT PradnyaParamita

[3] Darajati Wahyuningsih 2004 Strategi Pengelolaan Wilayah Pesisir dan Lautan SecaraTerpadu dan Berkelanjutan. Makalah dipresentasikan Sosialisasi Nasional MFCDP Jakarta

[4] Dahuri Rokhim 1998 Kebutuhan Riset untuk Mendukung Implementasi Pengelolaan Sumber daya Pesisir dan Lautan SecaraTerpadu *Jurnal Pesisir dan Lautan* 1. 2

[5] Pujayanti Adirini 2011 Budaya Maritim Geopolitik dan Tantangan Keamanan Indonesia Pusat Pengkajian Pengolahan Data dan Informasi Sekretariat Jenderal DPR RI

[6] Wahyudin Yudi 2003 Sistem Sosial Ekonomi dan Budaya Masyarakat Pesisir Disampaikan pada Pelatihan Pengelolaan Kawasan Konservasi Perairan Jakarta 5 Desember

[7] Desniarti 2003 Faktor Pendorong Perubahan Lingkungan Wilayah Pesisir Unpublish paper Institut Pertanian Bogor

[8] Dahuri R J Rais SP Ginting dan MJ Sitepu 2001 Pengelolaan Sumber Daya Wilayah Pesisir dan Lautan SecaraTerpadu Jakarta: PT Pradnya Paramita

[9] Lanza Randazzo. 2011. Improvements to a Coastal Management Plan in Sicily (Italy): New Approaches to borrow sediment management. *Journal of Coastal Research*, suppl. Special Issue 64 : 1357-1361.

[10] Zagonari, Fabio. 2012. An Optimization Model for Integrated Coastal Management: Development and a Case Study Using Italy's Comacchio Municipality. *Springer Science+Business Media* B.V..

[11] Helmick J S 2008 Port and Maritime Security: A Research Perspective. *Journal Transportation Security,* 115. 28.

[12] Sam Bateman. 2011. Solving the "Wicked Problems" of Maritime Security: Are Regional Forums up to the Task?. *Contemporary Southeast Asia.* 33, 1.

[13] Helmick J S 2008 Port and Maritime Security: A Research. *Perpective Journal Transportation Security.* 115.28.

[14] Casey, Allison, Matthew Sussex. 2012. Energy transit states and maritime security in the Malacca Strait: the case of Singapore. *Australian Journal of Maritime and Ocean Affairs.* 4.1.

[15] UNCLOS 1982 point 5

[16] Brotosusilo, Agus *et. All.* A Study on Socio-legal and Institutional Aspects of Coastal Resources Management in Segara Anakan, Cilacap, Indonesia." A Paper presented at the "US/ASEAN Technical Workshop on Integrated Tropical Coastal Zone Management." National University, Singapore, October 1988.

[17] Gomes, Takiyama. Pereira, Silva Ferreira. 2011. Social Diagnosis and Guidelines for Coastal Management in Environmental Protection Areas of the Amazon Littoral (Amapá, Brazil). *Journal of Coastal Research,* 64

[18] Helmick J S 2008 Port and Maritime Security: A Research Perpective Journal Transportation Security 115-28.

[19] Brotosusilo, Agus. Social Change in Segara Anakan. A Paper presented at the "US/ASEAN Technical Workshop on Integrated Tropical Coastal Zone Management." National University, Singapore, October 1988.

[20] Cordner, Lee. 2011. Progressing Maritime Security Cooperation in the Indian Ocean. *Naval War College Review.* 64. 4.

[21] Abdulrazaq, Abdulkadir, Sharifah Zubaidah Syed Abdul Kader. 2013. Port Security Vs Economic Gain: An Exposition of the Malaysian Port and Maritime Security Practice. *IIUM Law Journal.* 21. 2.

Environmental Awareness to Realizing Green Islamic Boarding School (Eco-Pesantren) In Indonesia

Herdis Herdiansyah[1], Trisasono Jokopitoyo[2], Ahmad Munir[3]
[1] Graduate Program, Universitas Indonesia, Salemba, Indonesia
[2]National Resilience Studies, Graduate Program, Universitas Indonesia, Salemba, Indonesia
[3] Environmental Studies, Graduate Program, Universitas Indonesia, Salemba, Indonesia

Emai: herdis@ui.ac.id

Abstract: This paper is based on Community Engagement Grants program which developed through stakeholder cooperation method. It means the involvement and participation of Pesantren community and the peoples around it as well as the institutions whose have aim related with the program try to realizegood environmental health for communities around it. Through empowerment approach which positioning Pesantren community, local people, and related institutionas a subject, the moral and religion values in Pesantren into moral value of environmental sensein form of value and knowledge that should be improved for the better environment health. Environmental development engagement could be done through two kinds of approaches. First is by project approach and the second is motivation approach or it could be both of them in integrated way. The second approach (motivation), although relatively spend longer time, it will effected more positive because the target will gradually change their attitude and behavior in persuasive manner. The behavior of ignoring environmental problem will turn into dynamically attitudes that will always grow and culminate in stability of environmental engagement

Keywords: environmental education, Islamic education, environmental awareness, *pesantren,* Engagement

1. Introduction

Environmental crisis in the modern society is caused by environmental management that ignored the sustainable principle. The damage of environment signed by environment degradation which makes modern civilization faces a condition to invade limited natural resources. Many methods have been done to solve this natural damage problem. However until now, there is no instant and quick formula to fix natural damage. Therefore, it needs another long term formula in solving this natural problem. One way that can be used is by educational approach [1]

In developing countries, environmental educational approach begins to be conducted massively. According to Bishnu's research, the environmental care approach in Asia Pacific is engaging in early time since this area has incredible natural resources but the damage area is also wide [2]. For example in Indonesia, environmental education is developed in formal school at early time. However since long time ago, there is traditionally educational system in Indonesia called Islamic Boarding School (Pesantren) as education institution that can be a key of educational development with environmental care basis.

Indonesia is the largest Muslim country in the world with more than 200 million adherents Muslims. There are more than 50.000 Islamic schools in Indonesia, where its religious lessons have direct and long term impactsfor Islamic living orientation inside this country.Islamic lesson began in

Indonesia since about the 13th century in form of Al Quranstudy at the village mosque, prayer houses, and at religion teacher's house. Islamic lesson was taught for children from six to eleven years old. During the time, various types of Islamic schools appeared in many regions of Indonesia, such as Pesantren and Pondok (mostly in Java and Kalimantan), Surau or mosque (in West Sumatra), Dayah (in Aceh), Madrasah and Islamic schools. *Pesantren*, *Pondok*, *Surau*, and *Dayah*were generally regarded as "traditional" Islamic educational institutions which have long history since Islam coming in Indonesia [3]

In Indonesia, Pesantren (Islamic Boarding School) especially traditional Pesantren is one of the traditional educational institution that still existed since hundreds years ago. In the beginning, Pesantren was a place to teach just about Islam, but now Pesantren has been transforming to become educational institution that not only teaches religion and morals, but also graduate cadres in economy, culture, and even environment sectors. In occasion of environmental sector in Indonesia, Pesantren which implementing green principles especially for environmental health has been grown recently. The Environmental health becomes crucial issue in creating eco Pesantren, since the former stereotype of Pesantren (especially traditional one) was seedy, squalid, and dirty. This perspective is not always wrong because many Pesantrens have less care of their students and workers' health due to the reason of isolated location, limited land, and limited funding. Therefore it negatively influences the health of traditional Islamic boarding school community [4]. Pesantren is assumed un-civilized as various diseases caused by dirty Pesantren'senvironment.

The special thing on boarding school education is learning from classical Islamic intellectual roommates treasure known as the yellow book (yellow book). In the past time, usually boarding school taught religion subject base, but now it is mostly completed with base elementary school subjects. This was the result of state Reformation in the late of 1970s. Besides non-religious subjects, many Pesantren also offer vocational courses such as agricultural skills, vehicle repair, and business enterprising skills, and environment awareness [5]. However mostly Pesantren has bad condition like badly need water, sanitation and hygiene, and the level of intensity of the relationship between human contacts is very vulnerable, especially in children particularly at risk to environmental health hazards.

2. Theoretical Overview

Environment degradation as environment social problem which is combination of supply, accessibility, and novelty of natural resources value that confirm the level of needed natural conservation. The stressed is environment value which is proportionally with the resource of natural resources that human needs. In this perspective, the core of degradation definition is externality, stack consumption, and error of differentiating human capital with natural capital. This perspective is dominated by relationship between consumer and reduce environment loss through scarcity natural resources consumption management. Natural degradation from consumption which consist of pollution and loss of diversity is indirect effects (Hampel, 1996).

The weakness of natural care in case of baddly environment management is happened since environment still assume as load. There are natural characteristics which caused difficulties for policy to internalise natural aspect into their policy. Those characteristics are:

1. Environment is intangible which means difficult to quantify.
2. Natural impacts will be happened in long term impact.
3. Natual impact will have characteristic of negative externality.
4. Environment as space and natural resources is assumed as public belonging.

However on Pesantren circumstances, the needed of synergized with environment is a must. Eventhough a lot of Pesantren are not able yet to realize its institution as educational institution with environmental friendly. Pesantren is expected to maintain and preserve the environment through Eco-Pesantren. The Society in this classtends to be apathetic to the environment. Commonly, the people will be aware after facing a greater impact due to environment damage. This is absolutely unwell because it will disturb people's activities and environmental conditions.Therefore, the engagement of maintaining the nature for public interest becomes together responsibility that cannot be negotiable anymore [6].

Besides improving religious knowledge, Pesantren Also teach the students to become useful people and could implement the values of the environment. Pesantren is mostly located in rural areas where the Kiyai, religious teachers, students, and administrators are lived together in one environment and based on their own Islamic values, norms, and habits. That makes this institution has a strategic position in carrying out the roles of education and socio-economic development for the community.

Those become effort to preserve the environment and increase the important of health awareness. In keeping Pesantrenenvironment well, some activities can be done such as recycling used goods, composting process, and applying technology for environmental pollution mitigation such as water treatment. Meanwhile in occasion of health, Pesantren can set uplocal herbal pharmacy by cultivating medical plants and all are arranged by *santri* or *santriwati* (Pesantren's student in both, male and female) under supervision of Pesantren staff.

As we know together that the natural condition now is dramatic and dangerous for the human health as pollution in the air, sea, and land. In addition, the climates change the caused damage and environmental pollution on earth roommates Showed by the increasing of floods, landslides and Drought. It is the caused by the deforestation where 5100 growth natural forests were lost every day.

Islam clearly gives instruction to maintain the nature and its balance. Moreover, Islam also warns us for not to break and damage the environment [7]. The environment that was beautiful in the former, is destroyed by irresponsible hands recently. This is caused by humanitarian and social morality crisis as well as human lifestyle that tend to be consumptive who result in loss of natural balance [8].

Many results of study about environment care effectiveness shows that the efforts of natural conservation from a kind of institution will create a number of benefits especially for its community and people in general. Hence, through education, environment care could be continually developed. Study of Larijani (2008) shows that when a teacher could deliver their science and experience in environment management, the student will gradually follow and practice it in daily life and in their environment [9].

3. Methodology

This activityis carried out through community approach as a subject. The methods of this research is based on the result of community engagement grants. In addition, statistical assessment by sustainable index methods is also conducted. The Ecopesantren sustainable index uses Multi-Dimensional Scaling (MDS) approach. This concept is actually modifying the approach of Rapid Appraisal Techniques for Fishery (RAPFISH) at the University of British Columbia (Kavanagh, 2001) where initially measures the sustainability of resources and fish catches.

Pesantren shares its value and knowledgeas well as develops and improves it to be better than before. It has conducted community empowerment activity through Eco-Pesantren education as the basis of moral character building with environmental sense. The activities are :

1. The establishment of PesantrenWaste Bank (3R, reuse-reduce-recycle)
   One of the problems that many people who live in urban or rural areas faces is wastemanagement.The greater waste sources are coming from household, agriculture, industry, service, and office activities. That waste have potentially polluted and damaged the environment instead natural resources. Meanwhile waste management is actually very easy by process of recycling for instance.
2. The Simple Waste Water Installation (Sanitation and Clean Water)
   Pesantren area should be produced a lot of waste water without proper treatment and gradually impacted on environment pollution. Therefore, the water treatment can be used to treat unused water to be used again for public interest.
3. Green Pesantren and Land Exploration
   The application of Green Pesantren program is by planting various plantations such as ornamental plants, vegetables, fruits, or herbal plants. The planting uses several methods such as vertical garden, hydroponic, green house, and this circumstance will beautify Pesantren itself.

4. Green Curriculum in Teaching Process Includes Library and Information Service with Environment Basis.
   Green School Movement should be arranged holistically by linking the whole programs in Pesantren and consideringvarious influenced factors such as supporting or inhibiting factors. The Internal potentialPesantrenare in form of land, water resources, energy, and waste as well as local wisdom as society ones around Pesantren.

4. Result

According to Rapfish, the modification of techniques that adjusted with sustainability of Ecopesantren is done. Method of MDS is statistical analysis technique with computer basis by using statistical software on computer. It performs transformation of any dimension and multidimensional of renewable resources sustainability management in border area. Data analysis with MDS covers sustainability aspect on dimensions of ecological, economic, and social culture.

Table 1. Sustainability Index and Statistical Value

Dimension	**Sustainability index**		**difference**	**category**	**Statistical value**		**Iteration**
	MDS	Monte Carlo			Stress	R^2	
Multidimensional	62,64	62,66	0,02	sustainable enough	0,2147	0,8972	2
Ecology	50,20	50,22	0,02	sustainable enough	0,1532	0,942	3
Economy	64,58	64,56	0,02	sustainable enough	0,15135	0,9023	2
Social culture	53,05	53,04	0,01	sustainable enough	0,1349	0,9506	3

In appointing Ecopesantren sustainability index, one of the main keys to turn the concept into a measurable sustainability index conducted by MDS is finding or even input of quantitative data. The types and forms of data is highly dependent on character and type of conducted research. The index value of Ecopesantren sustainability status is determined on the scores for each attribute in each dimension in accordance with the conditions of recent management with reference to the criteria of sustainable development concept.

The result of research shows that Ecopesantren sustainability level is in the level of sustainable enough (50.01 to 75). This describes the condition of existing Ecopesantren is in a good level, but unsustainable yet.

In the progression, any Islamic boarding school has developed in various norm and ethic [10-11]. One of the examples is the Eco-Pesantren. Eco-Pesantren the model is arranged in order to the make boarding school as a center of environmental study based on Islamic science for the students and the community around it. The development with human ecology approach will mix the right boarding school as united social-cultural. This model of the make doubt of few parent, however, Kiyai will motivate the students that this could be a model of provision for reviews their future life.

Environmental motivational approach can be done in Pesantren education management. The sense of environmental balance that arises from base understanding of the problems and the implication of *ukhrawi* and worldly welfare can be taught and developed through education in this environment. Meanwhile the development of the green area and its supporting system is a part of environmental management program. The development of environmental basis curriculum can be integrated arranged or monolithic. Material development, learning models, and various learning methods are conducted to provide insight for the students and people around about environment which associated with everyday environmental local issue.

So many success stories with their own characteristics in every region, and now it is trying to be applied in Pesantren. Usually, waste is just throwing away without sorting or recycling process. Through the development of Waste Bank system, the students will be more motivated to sort the waste and then get moral lesson on how we deal with waste. If the Waste Bank system success in Pesantren, the local community would be interested to adopt it. According to Sabah research (2008), environment protection raised as passion of environment guard. The success of environment program can result in the willing of spirit reproduction in another form based on environment protection [12].

Hence, the Eco-Pesantrencommunity engagement as environmental character building basis is easy to be implemented at Pesantren and the people around it. The Indonesia's natural condition which degrade and danger for human health is caused by people's vandalism.The green Pesantren development through cooperation of stakeholders can be realized as principles and rules of environmentally education development. Pesantren as basis place for character and Islamic morality builder which synergized environment sense inside is demanded to comprehend not only religion and science subjects, but also becomes role model for the future humans who pro and care with the environment.

5. Conclusion

The results of activities showed that environmental sense and health could be optimized when it actualized in form of integrated environmental education curriculum. The good environment situation could be created through approaches of *knowing*(knowledge, cognitive), *being* (attitudes), and *doing* (psychomotor). In addition,the value and norm of environmental health, behaviors, water management for health standard, fresh ventilation and air circulation, waste management in 3R (Reuse, Reduce, Recycle), as well as available public facilities are also contributing good Pesantren environment.

The results of environmental health program show significance changes as follows:

1. The program is more optimum after actualized in the form of integrated environmental education curriculum. The good environment situation could be created through approaches of *knowing* (knowledge, cognitive), *being* (attitudes), and *doing* (psychomotor). In addition, the value and norm of environmental health, behaviors, water management for health standard, fresh ventilation and air circulation, waste management in 3R (Reuse, Reduce, Recycle), as well as available public facilities are also contributing good Pesantren environment.
2. The raise of Pesantren's efforts to protect and maintain health facilities that related with environment. The examples are establishment of regulation for Pesantren community to create environment order and cleanliness as well as punishment for the violence actor.
3. The cleanliness program which is held regularlyin daily or weekly and involved surrounding community as well as cleanliness competition between religious institutions.
4. The support from competent institution to conduct health surrounding communities.
5. The squalid, seedy, exclusive (closed), anti-socialization, and social un-responsibility images in Pesantren before become degrade and disappear through together environment cleanliness and health activitythat integrated and merged with surrounding community nowdays.

ACKNOWLEDGMENTS

This paper is national collaboration of community engagement and was supported with DRPM Universitas Indonesia and Ministry of environment and foresty of Indonesia

6. References

[1] Lincoln R. Larson Steven B. Castleberry Gary T. Green. 2010 Effects of an Environmental Education Program on the Environmental Orientations of Children from Different Gender, Age, and Ethnic Groups . *Journal of Park and Recreation Administration.* **28**,

[2] Bishnu B. Bhandari and Osamu Abe 2000 Environmental Education in the Asia-Pacific Region: Some Problems and Prospects *International Review for Environmental Strategies* **1**, 1, 57 – 77

[3] Kurniawati, Dewi. 2013. Darunnajah Islamic Boarding School: A Form of Moderate Islamic Teaching in Indonesia. London: Kantor Atase Pendidikan KBRI.

[4] Abu-Hola Imfadi 2009 An Islamic Perspective on Environmental Literacy*Education Journal* **130** 2 p 195

[5] Tan Charlene 2014 Educative Tradition and Islamic Schools in Indonesia. *Journal of Arabic and Islamic Studies***14** pp 47-62

[6] Maryam Larijani. Assessment of Environmental Awareness among Higher Primary School Teachers. J Hum Ecol, 31(2): 121-124 (2010)

[7] Boyd J M 1984 The Role of Religion in Conservation *The Environmentalist* **4** 7

[8] Sabah A. Abdul-Wahab 2008 A Preliminary Investigation into the Environmental Awareness of the Omani Public and their Willingness to Protect the Environmennt *American Journal of Environmental Sciences* **4** 1 pp 39-49

[9] Maryam Larijani. Assessment of Environmental Awareness among Higher Primary School Teachers. J Hum Ecol, 31(2): 121-124 (2010)

[10] [14] Burhanuddin Jajat and Jamhari 2006. Assessment of Social and Political Attitudes in Indonesia Islamic Education Institution. *Studia Islamika: Indonesian Journal for Islamic Studies,* **13** 3 pp 399-433

[11] Deuraseh N 2009. Maintaining a Healthy Environment: An Islamic Ethical Approach. *European Journal of Social Sciences* **8** 4

[12] Sabah A. Abdul-Wahab 2008 A Preliminary Investigation into the Environmental Awareness of the Omani Public and their Willingness to Protect the Environmennt *American Journal of Environmental Sciences* **4** 1 39-49

Sustainability of Water Resources in the Upstream Watershed-Based Community Engagement and Multistakeholder Cooperation

Agus Brotosusilo[1], Dyah Utari [2], Afrizal Agung Satria[3]
[1] Faculty of Law, Universitas Indonesia, Depok, Indonesia
[2] Faculty of Public Health Universitas Indonesia, Depok, Indonesia
[3] Graduarted Program Universitas Indonesia, Depok, Indonesia

Emai: broto.susilo@ui.ac.id, agus_brotosusilo@yahoo.com

Abstract: The communities engagement become the backbone of the conservation in the Citanduy upstream watershed. It functioning as a major deal and the first one in keeping his own Watershed. This paper based on Community Engagement Grants (CEGs). Program Society-based empowerment approach is also emphasized in the viewpoint of environmental law that is useful to set governance and sanctions in watershed management. The type of activity to be undertaken are the expansion of awareness programs communities of the existence and condition of the watershed Citanduy, the formation of a cadre of conservationists environment that is primarily directed to children and women, the institutionalization of customary law environment, and afforestation by planting 100,000 prolific trees, tree conservationists, and Sunda endemic tree in the land surrounding the watershed upstream Citanduy. The Program involves several partners and stakeholders who helped in substance and operational support activities in the field.. Result of program shows that Community Engagement Grants need cooperation among stakeholders by positioning the community as main subject of changing, not as subject who does not understand their needs to change.

Keywords: Upstream Watershed, water security, Community Engagement, Multistakeholder Cooperation

1. Introduction

The existence of natural resources that can be renewable, especially water, is essential for humans, because humans are very dependence of the water in their life sustainability cycle, that people in this earth and on other planets can not live without it. Much more water, which is a natural resource that man at all times, whether for drinking, food ingredients, Eating Wash latrine (MCK), and others [1]. The growth of world population is increasing and the demand for water is increasing push more water than the strengthening of the economic value and social function. The condition has the potential to create conflicts of interest (horizontal and vertical) between sectors, between regions and the various parties related to water resources. To that end, the continuity and sustainability of clean water supply must be well managed and cared for by all sides, including the community, to realize the management of water resources in an integrated sustainable by always fulfill the function of the environment and economy in harmony and maintain a balance between the ecosystem and power environmental support, as mandated by Law No. 7 of 2004 on Water Resources (Law 7/2004). Conflict between groups may be more massive because water resources is one resource capture vital [2] .

According to data from the Council of Water Resources (2014), the amount of water in the world now around 1.404377 billion km3 and the amount of fresh water only reached 33,705,048 km3 (2.4% of total world water) as well as the number of 131 460 km3 of surface water (0.009% of the amount of water in the world). This shows that the conditions for protecting and preserving the water supply has been serious, because the amount of water in the world amounted to only 2.4 percent. This condition raises awareness of the importance of water security that is related with certainty that everyone has easy access to clean water and affordable price in order to achieve a healthy life, decent and productive [3]. Water security is mentioned UNESCO [4] are measures to meet the need for clean water and the need to obtain a sufficient supply of clean water. The security of water conditions in line with the fulfillment of the right of every human being to have access to clean water and inadequate fulfillment of the needs of clean water so that in order to fulfill the need for clean water, some efforts for the preservation of clean water continue to be done.

The problem of water security is also happened Citanduy Watershed which extends from West Java to Central Java and covering 10 districts/cities at Ciamis, Tasikmalaya, and Cilacap. Citanduy watershed consists of critical land in very critical level for about 2460.88 ha (0 , 67%), critical level for 14793.38 ha (4.05%), potential critical for 117,730.22 ha (32.27%), quite critical for 63911.16 ha (17.52%), and not critical for about 165.977,10 ha (45.49%). Citanduy waterhed is one of 22 watersheds in Indonesia that are classified critically since facing various ecological problems such as erosion and sedimentation which causing flood, the reduction of area, the decrease of lagoons depth, river pollution, environmental degradation quality and other related problems. The 2013 Ministry of Public Works data stated that the Citanduy watershed have erosion level which categorized as very danger is 26.764.31 ha (7.34%); categorized as danger level is 37.499.02 ha (10.28%) .

The land erosion that was being wider is 104,927.13 ha (28.74%), and the smaller erosion level is 195,682.27 ha (53.63%). The total erosion is 28,962,668.00 tons / year. This condition is exacerbated by the landslide level of Citanduy watershed which can be classified as highly vulnerable of 2,639,80 ha, and susceptible of 6.501.35 ha. For the data of flood level, the area that is very vulnerable is 15.554.72 ha (4.26%), the area that potentially flood is 129,053.72 ha (35.37%), the area that somewhat prone to flood is 3.735.43 ha (1,02%), and for about 216,528.88 ha (59.34%) is not prone to flooding. The degradation of Citanduy watershed condition has threatened the availability of clean water and even drinking water for people around watershed because in the dry season there will be a drought.

If we are referring to water resource potential, it is influenced by its physically characteristic which consist of various type and formation of stone, relief or topography, type of soil, and land use. Those characteristics will influence water resource potential that can be seen from its differences of quantity and quality in each area [5]. The potential water resource can be seen from its quality of water surface and spring water potencies. The water surface potencies are river, pond, lake, reservoir, and dam. The area of Citanduy river fed by 17 branches of river. The understanding of positioning community as actor of changing is based on the fact that the former community engagement grants were positioned community as object, not a subject. On the other hand, this paper will be exploring community to become actor of changing.

2. Theoretical review

Management of natural resources and integrated sustainable water security in line with the definition promoted by UNESCO [6]. Water security challenges facing complex and interrelated and become a major theme of water issues in order to achieve a security, sustainability, and feasibility for humans both locally and internationally. When water security conditions are not going well or when access to water is disturbed, people face acute human security risks that are connected with poor health and bad life.

Citanduy River region is one of the Cross River Basin Province, located in the province of West Java and Central Java the south in accordance with the Decree of the President of the Republic of

Indonesia Number 12 Year 2012 on the Establishment of the River Basin (02:10 WS Code A2). Citanduy watershed is one of the priority watersheds in Java. Priority watershed sedimentation is caused by a very large part of the river downstream Citanduy. Geographically Citanduy river area located at position 1080 04 'until 1090 30' east longitude (BT) and 70 03 'to 70 52' South latitude (LS). Citanduy DAS consists of six (6) Sub, namely Citanduy upstream subzone, subzone Cijolang, Cikawung subzone, subzone Cimuntur, Ciseel subzone, subzone Citanduy Upper and subzone Segara Branches.

Critical lands, erosion and flood-prone is increasing from year to year so that the necessary government efforts and public participation in the environment to not aggravate land degradation and improve watershed Citanduy back role as a source of clean water supply providers. Community activities around the watershed Citanduy directly and indirectly affect water quality, sustainability of water supply and environmental conditions Citanduy watershed. During this watershed Citanduy into the sand mining areas, use of forests for the surrounding community activities, community activities such as for irrigation and household activities. Community activities in the utilization of this water has exceeded the carrying capacity of the environment surrounding the watershed Citanduy characterized by the onset of flooding in the rainy season and the river drying up in times of drought.

Besides reducing floods, droughts, and landslides, watershed conservation will bring multi-effects. Therefore, water conservation should get great attention. Currently, the water conservation in the city becomes one of the main keys to ensure availability and increasing of water supply in accordance with water demand [7]. The society needs to have knowledge about condition of riparian zones which is weak inherently ecosystems. The urban development activity is necessary in order to avoid increasing of water temperature and light intensity which are very disturbing the watershed natural ecosystems [8].

Ellis [9] showed the most important element of society in conservation is at household level. Conditions that affecting the involvement of citizens at household level are due to endogenous factors such as relationships/relations in social, institution, and organization. Meanwhile the exogenous factors consist of trend and policy as well as threat/interference. That condition needs to be considered because providing access for society to be able to participate in various conservation activities, including watershed are in the community. In order to conducting conservation or watershed management, there are two approaches which related to community-based activities and society participation. Those are: 1) rational choice approach, where the person or group will behave in certain ways since logic of appropriateness engagement as the result of cognitive understanding and internalization process that takes relatively long time and a "groupthink" is formed. Meanwhile, 2) in normative approach, the behavior of person or group is a response of rules or incentives, where the rules can be formal rules in form of Laws, as well as non-formal rules from society agreement [10]. The involvement of stakeholders is stated by Williams [11] that " allows local stakeholders to identify politically and socially acceptable management practices". This collaboration is contribute concrete step and could be operated effectively. Technically, the stakeholders will also prepare needs at the location of can be liaison from servant team to the society as participant of program.

Watershed conservation is very important since river is clean water resource asset for drinking and other interests, ecological conditions as source of animal protein, as well as social culture and local wisdom. Those are ecological values of riparian area that are important to be managed for watershed conservation, where the nature which has river vegetation could be helpful for Keeping the water quality and clarity by reducing sediment and run-off nutrient from land, reducing the level of flood by increasing water retention time (but not to the level of river dried when dry season), keeping low water temperature by providing deposit, energy affects variation by influencing light quality and terrestrial organic matters input, providing habitat for aquatic invertebrates and fish (for instance are in the form of cover of leaves and woody debris accumulation), and for land plants as well as animals, providing breeding area for native birds (for instance blue duck) and fish (for example anchovies, providing suitable corridor for native plant and animal extension as well as space for high species diversity, and improving soil stability and minimizing cliff erosion at the river [12].

3. Methodology

The location of program is at Cikadu village, Guranteng, Pagerageung district, Tasikmalaya regency, and at Pasir Jaya village, Buana Mekar, Panumbangan district. Those area are placed side by side at the Citanduy watershed. Both locations are located about 260 km from Jakarta city. The locations are chosen because especially for Cikadu village, it is location of Community-Based Services of the University of Indonesia before (November 2014-June 2015). Several stakeholders were synergized with that program which results in positive changing and impact for Cikadu village.

The methodology of activities are conducted through some stages, those are:

a. Problem identifications in the field which are related with activities of the people and the water resources at Citanduy watershed.
b. Knowledge sharing by transfer of knowledge and success story of watershed management at other places. It also shares about the problem and the people urgency to participate with Citanduy watershed conservation accordance with the perspective of environmental laws.
c. Program implementations are in form of activities such as tree planting in accordance with Citanduy watershed vegetation, simple technology water management, and other conservation actions. This step is replicating the PROKASIH program (Clean River Program) and SUPERKASIH program (Statement Letter of Clean River Program) from the government which involving not only the industry but also the people.
d. Monitoring the success and sustainability of programs by supervising how far the participants followed the program and when they back to their environment, they can maintain the clean water sustainability based on empirical experience during the training.

These activities are supported by the Provincial Government of West Java (BPLHD West Java), Assistant Deputy of Public Organization Role Engagement from Ministry of Environment and Forestry, Department of Plantation and Forestry of Tasikmalaya regency, Department of Plantation and Forestry of Ciamis regency, and "Geliat" as local NGOs who became partner in the field. Each of them has contribution in this program. The program objectives are giving socialization, training, simple and applicative watershed management to the people through environmental law perspective for clean water sustainability program at Citanduy watershed. The society becomes important stakeholder in this case since they are the closest part of watershed area, most interest, and have great potential for being agent of changes.

4. Result and discussions

The Citanduy watershed topography is in form of mountains at the Northern and mainland at the Southern which bordering with Nusakambangan island. According to its topography, Citanduy watershed divided into 3 parts, namely the upstream watershed, middle watershed, and downstream watershed. Understanding and participation of the community to maintain the sustainability of natural resources are needed in order to maintain watershed functions Citanduy as water providers in the region of West Java and Central Java. So far the Government has made various efforts such as Critical Land Rehabilitation Movement (GRLK), National Movement for Forest and Land Rehabilitation (GNRHL / GERHAN), as well as reforestation activities to address environmental problems in the watershed Citanduy. But the GOI's effort has not been optimal so that the community engagement efforts must be continued. Responsibility for the preservation of the watershed Citanduy can not be completely handed over to the government, because without the participation of the community then no matter how much development efforts and improvements made by the government will not solve the problems Citanduy critical watershed lands [12-13].

Community involvement should be based on the awareness of society to contribute to maintain the stability of ecosystems in the watershed Citanduy and not do activities that cause environmental and river damage. In addition, community involvement is done by taking into account local knowledge as recognized by Law No. 32 of 2009 on the Protection and Management of the Environment (Law 32/2009).

Table 1. List of Changing of Community Engagement in Location

No	Condition before	Activities	Condition after
1	Sedimentation threat cause of unmeasured logging.	It has been planting almost 30 thousand trees fixed with watershed characteristic.	The empty area is planted with various trees nowdays.
2	Low of community careness in downstream watershed conservation	Cooperation program which involving community is hold.	The sustainable *gotong royong* (working together) is happened.
3	No penalties for environment broker actor	Local regulation based on social culture of local community is established.	The people begin to follow the environment laws that were made together.

Related to the number of community activities that damage the environment in the watershed Citanduy, we need to hold social interventions in the form of environmental law enforcement efforts to deal with these problems and training and socialization of public participation to environmental sustainability in watershed Citanduy. Enforcement of environmental law is based on Law 32/2009 taking into account the principle of ultimum remedium namely the application of criminal law enforcement as a last resort after the application of administrative law enforcement is considered unsuccessful.

When people are able to participate actively in clean water sustainability program, the task of government will be easier and the state budget will be more efficient for watershed normalization project. On the other hand, the failure of community involvement will make people apathetic with the watershed and potentially damage the watershed. It will result in decreasing of clean water supply and finally becoming problems of the government to fix and fulfill that supply.

5. Conclusion

The expected results are the growing of people's understanding and concrete action of conservation management about the important of people's participation in supporting clean water supply at Citanduy watershed. The surrounding society is an important stakeholder for this purpose, because they are the closest part with the watershed, the most dependent, and have great potential as agents of change. When people are able to participate actively in clean water sustainability program, the task of government will be easier and the state budget will be more efficient for watershed normalization project. On the other hand, the failure of community involvement will make people apathetic with the watershed and potentially damage the watershed. It will result in decreasing of clean water supply and finally becoming problems of the government to fix and fulfill that supply. Moreover, the involvement of related stakeholders is obligatory for the success of program since its characteristic of supporting and cooperating each other.

6. References

[1] Ellis F 2002 *Rural Livelihoods and Diversity in Developing Countries.* Oxford: Oxford University Press

[2] Grecksch Kevin 2015 Adaptive capacity and water governance in the Keiskamma River Catchment, Eastern Cape Province, South Africa. *Water S.A* 41.3: 359-368

[3] Gautier Catherine 2008 Oil Water and Climate: An Introduction; University of California, Santa Barbara

[4] UNESCO - UN Water 2014 Analytical Brief The World Water Development Report 2014

[5] Brotosusilo Agus 1988 Social Change in Segara Anakan. Presented paper at the "US/ASEAN Technical Workshop on Integrated Tropical Coastal Zone Management", Singapore: National University.
[6] UNESCO - UN Water 2014 Analytical Brief: The World Water Development Report 2014
[7] Asteria Donna and Agus Brotosusilo 2015 Model Optimalisasi Partisipasi Warga Perempuan Sebagai Modal Sosial dan Agen Perubahan dalam Konservasi DAS di Perkotaan Secara Berkelanjutan. Depok: DRPM UI
[8] Asteria Donna and Agus Brotosusilo 2015.Model Optimalisasi Partisipasi Warga Perempuan Sebagai Modal Sosial dan Agen Perubahan dalam Konservasi DAS di Perkotaan Secara Berkelanjutan. Depok: DRPM UI
[9] Ellis F 2002 Rural Livelihoods and Diversity in Developing Countries Oxford University Press Oxford
[10] Asteria Donna and Agus Brotosusilo 2015 Model Optimalisasi Partisipasi Warga Perempuan Sebagai Modal Sosial dan Agen Perubahan dalam Konservasi DAS di Perkotaan Secara Berkelanjutan. Depok: DRPM UI
[11] Williams J.R, Smith, C.M, Roe, J.D., Leatherman, J.C., & Wilson, R.M 2005 Engaging Watershed Stakeholders for Cost-Effective Environmental Management Planning with "Watershed Manager". *Journal of Natural Resources and Life Sciences Education,* 41, 1.
[12] Brotosusilo Agus 1988 Public Officials Attitude Pertaining to Coastal Resources Management Plan Presented paper at "US/ASEAN Workshop on Marine Sciences". Semarang.
[13] O'Brien M 1993. Community Perspective of Riparian Management and Restoration: A case study in Marlborough, in Collier K.J (Ed)1994. Restoration of Aquatic Habitats (selected papers from the second day of the New Zealand Limnological Society 1993 Annual Conference). Department of Conservation, USA, pages 145-162.